D0372836

That's Not in My Science Book

That's Not in My Science Book

A Compilation of Little-Known Facts

Kate Kelly

Taylor Trade Publishing

Lanham • New York • Boulder • Toronto • Plymouth, UK

Copyright © 2006 by Kate Kelly
First Taylor Trade Publishing edition 2006

This Taylor Trade Publishing paperback edition of *That's Not in My Science Book* is an original publication. It is published by arrangement with the author.

Detail of nanogear diagram on page 209 courtesy of NASA.

All rights reserved. No part of this book may be reproduced in any form or by any electronic or mechanical means, including information storage and retrieval systems, without written permission from the publisher, except by a reviewer who may quote passages in a review.

Published by Taylor Trade Publishing
An imprint of The Rowman & Littlefield Publishing Group, Inc.
4501 Forbes Boulevard, Suite 200, Lanham, Maryland 20706

Estover Road
Plymouth PL6 7PY
United Kingdom

Distributed by NATIONAL BOOK NETWORK

Library of Congress Cataloging-in-Publication Data

Kelly, Kate, 1950–
That's not in my science book : a compilation of little-known facts / Kate Kelly.
p. cm.
Includes index.
ISBN-13: 978-1-58979-290-6 (pbk. : alk. paper)
ISBN-10: 1-58979-290-4 (pbk. : alk. paper)
1. Science—Popular works. 2. Discoveries in science—Popular works. I. Title.
Q162.K44 2006
500—dc22

2006009129

The paper used in this publication meets the minimum requirements of American National Standard for Information Sciences—Permanence of Paper for Printed Library Materials, ANSI/NISO Z39.48-1992.

Manufactured in the United States of America.

This book is dedicated to fellow writer Marian Edelman Borden. When I called her with news of this book project, she congratulated me and then said, "Wow. The writer who wrote *That's Not in My American History Book* only had to be responsible for three hundred years. You've got to cover all of eternity."

And then she supported me every step of the way.
Thank you, Marian.

Contents

Part Three

Amazing Discoveries That Changed Our View of the Universe

Part Four

Protecting Our Living Planet

Part Five

A Peek at the Future

Acknowledgments

The challenge of this book was to make science fun and accessible, and just as scientists stand on the shoulders of those who precede them, writers, too, benefit from others who have wrestled to make difficult concepts understandable. I particularly want to acknowledge the scientific background I gained from reading and listening to work by Robert M. Hazen, PhD, a professor of George Mason University, who also is on the staff of the Geophysical Lab at Carnegie Institution of Washington. Together with James Trefil, coauthor, physicist, and professor at George Mason, Hazen has provided a wonderful service by taking a wide range of scientific material and teaching us about it in a sensible, practical manner. Hazen also created *The Joy of Science*, a sixty-part lecture series (in audio and video versions). For difficult subjects like magnetic stripping and seafloor spreading (see chapter 2, "The Earth Moves in Mysterious Ways: Plate Tectonics, Earthquakes, and More"), it was wonderful to be able to pop one of his lectures into my VCR and play it several times until I fully understood how I could convey this information to readers. My intent has been to further translate challenging material, but for those who are ready for a slightly deeper understanding of a broad array of science subjects, I recommend Hazen and Trefil's *Science Matters: Achieving Science Literacy* as well as Hazen's *Joy of Science* course.

Bill Bryson's *A Short History of Nearly Everything* was also inspirational. Bryson brings a story component to every topic he tackles—a wonderful way to gain a deeper understanding of science.

Writing *That's Not in My Science Book* gave me an opportu-

nity to pay close attention to a couple of different organizations that were important in helping me shape the book. At a meeting of the Wildlife Trust, a New York–based international organization ably led by Dr. Mary Pearl, I had occasion to meet one of the board members, Pamela Thye, who introduced me to a working committee, the Wildlife Health Sciences Committee, at the Wildlife Conservation Society (the Bronx Zoo). While we may think of the Wildlife Conservation Society as "the zoo," it actually has a huge outreach and is active in international conservation of both wildlife and the natural wild lands that are needed to support the animals. The WCS has programs in Africa, Asia, and Latin America as well as North America. The Wildlife Health Sciences Committee, led by Dr. Robert A. Cook, chief veterinarian of the WCS, and Dr. William Karesh, who heads the field vet program, are leading an effort to implement a formal system to surveil both human and wildlife health. They can document that our more densely settled world is causing diseases formerly restricted to wildlife to transfer to the human population. (See chapter 9, "Avian Flu: A Dangerous Virus and Its Long Shadow.")

I need to back up for a moment to thank the Wildlife Trust staff as well. Mary Pearl and her organization participated in the creation of the Consortium for Conservation Medicine, with the intent of combining efforts among scientists and medical personnel to interrupt disease transmission routes to prevent or decrease future outbreaks. And thanks to Linda Shockley of the Trust, who was very responsive to me. I appreciate the help.

The American Museum of Natural History is the repository of so much information about our world, and it provided background for many of the chapters. (While the entire museum is a treat, anyone who has not been to the dinosaur exhibit is missing out.) And Dr. Marc W. Buie, an astronomer at the Lowell Observatory, was very helpful to me in coming to a better understanding of the "is it or isn't it a planet?" question regarding Pluto.

When it came to selecting topics, I had wonderful advice from fellow writer, Marian Edelman Borden, who helped shape the book, and Dr. Hannah Shear, a research immunologist who has opted to share her knowledge through teaching. She, too, had helpful suggestions as to what topics I should cover. Dr. Ann Engelland and Ardis Danon both shared with me their ideas for chapter 6, "How They Learned Why We Get Sick: The Origin of Germ Theory." Teachers John Elia, who teaches at New Rochelle (New York) High School, and Christopher Ward, who teaches at Hommocks Middle School in Larchmont, New York, and former science professor John C. Greene were all willing to give their time and attention to helping me provide accurate information in an understandable way.

To be given the opportunity to dive into such a wide range of fascinating topics was one of the greatest gifts—and challenges—I could have received. I will always be grateful to fellow writer George Sheldon, who was kind enough to alert several of us to the possibility of this project, and I thank agent Robert DiForio for representing me through the process.

At Taylor Publishing, Rick Rinehart was patient while I continued "on task," and the book benefited from the copyediting of Patricia Zylius as well as the attentive work of production editor Jehanne Schweitzer. Thank you.

My final sentiment is utmost respect and a deep sense of gratitude for all the men and women who have dedicated their lives to scientific investigation and have pushed the boundaries of what we know to make our world better. If this book inspires even one more person to go into the field, then that's very exciting.

Introduction

Before you start reading the chapters, I want to reassure you that this is a book for the layperson. Your tour guide here is someone with an intense curiosity and a high-level need to make sure everything is understandable—and fun. The book is filled with amazing stories, with a few jokes thrown in to lighten things up.

I knew from the beginning that being asked to write *That's Not in My Science Book* was both a gift and a dilemma. The assignment was a gift because I was given permission to investigate anything that interested me—and quite a few things that I knew would be good for me to cover—but it presented a terrible dilemma because, inevitably, the book must be inadequate. How could I even begin to do justice to the universe in only a few hundred pages?

I write this introduction at the end of what has been a fascinating journey. I've tried to cover a fair number of significant developments in science, so you'll find chapters on Copernicus, atomic theory, Einstein, and evolution—topics that simply couldn't be left out. But I've highlighted the kinds of details that are a little off-topic for a textbook. Other subjects were included because "they would never be in a science text book." Once you've left elementary school, you don't have time to study elephants, and few people ever have the opportunity to read about the beginnings of our environmental movement. As for dinosaurs, they are generally left to the under-ten set. Other subjects were covered because inquiring minds (mine) wanted to know "Who thought of *that*?"

I leave mounds and mounds of research untouched because

there wasn't space. Had I pursued every subject that interested me, the book would have become a weighty tome, not the quick and pleasurable romp through the world of science that it is intended to be.

Some Basic Conclusions

As I ranged from topic to topic, some general themes kept recurring, and I'll share them with you now.

- The Greeks were an amazing people. As you will see with virtually every subject, the Greeks had a lot of extremely advanced concepts pretty much nailed down. Our civilization would be very different if we had been able to build on their theories earlier. (So much discovery had to wait until the eighteenth and nineteenth centuries.)
- "If I have been able to see further, it was only because I stood on the shoulders of giants," is a quote from Sir Isaac Newton, who says so well that every discovery in science is based on the amazing work that preceded it. Science is about the gradual clarification and elaboration of ideas, with contributions from many hands. Newtonian concepts of space, time, and gravitation served their purpose from 1687 until now. In the twentieth century Einstein came along and developed even more truths in the field of physics.
- In science, a theory describes a body of knowledge that has stood the test of time and observation. "It's only a theory," is an argument sometimes used against scientific ideas, but in fact, for something to qualify as a scientific theory it means a hypothesis (an educated guess) has actually been tested, refined, and proven to the degree possible for the day.
- While science is all about proof, there is a fair element of "You just gotta believe . . ." Though wonderful advances were made in chemistry throughout the nineteenth century, there was no real proof of the existence of atoms until 1905 (Einstein and Brownian motion), and atoms were not successfully photographed until 1980. Imagine all that

would have been lost if scientists had insisted on actual proof before continuing their work!

- The scientists you read about here are mostly white men. While diversity is finally entering the world of science, until recently there has not been much opportunity for anyone other than white men to "join the club." You'll find that a few women are belatedly being acknowledged for some major accomplishments, but for the most part, the opportunities for women and people of different ethnicities were extremely limited.
- Virtually no area of science is free of doubt or gaps in understanding, and scientists are never "done." While Einstein is certainly the Mick Jagger of scientists, even his work is now being used as a launching pad for more advanced thinking.
- It is perfectly normal for the world to be slow at recognizing the importance of a new discovery. As in all fields of knowledge, it is sometimes difficult to see what is right before your eyes. If Darwin and his contemporaries had only fully realized the significance of the bird-reptile specimen (the *Archaeopteryx*; see chapters 3 and 7) that was brought into the British Museum of Natural History only two years after *The Origin of Species* was published, they would have had quite a field day.

Why It's Important That Books Like This Are Read

And finally, I want to share with you a quote from Louis Pasteur: "Did you ever observe to whom the accidents happen? Chance favors only the prepared mind." He, too, so well depicts the openness necessary to discovery anything in science—or in life.

Our world is changing every day, and many of these changes are dictated by science: Some changes are the result of new technology (stem cell research, space flight), and others are the result of progress and what happens to our world as a result of it (global warming, pollution). Because we are fortunate enough to live in a democracy, our opinions actually matter. For that reason, we need some basic understanding of science so that we can develop educated opinions about what is happening.

In addition, scientific knowledge in the field of medicine and nutrition has increased exponentially, and as a result, a little knowledge of the science behind medical treatments or nutrition guidelines can make us better consumers—and healthier people.

While scientific literacy has improved over the past couple of decades, Dr. Jon D. Miller, a political scientist who directs the Center for Biomedical Communications at Northwestern University Medical School, has conducted numerous surveys regarding science literacy. In an interview with the *New York Times* (August 30, 2005) he reports that only 20–25 percent of Americans are "scientifically savvy." Dr. Miller's research reflects some major gaps in basic knowledge: American adults in general do not understand what molecules are; fewer than a third know that DNA is a key to heredity; only about 10 percent know what radiation is; and one adult American in five thinks the sun revolves around the earth—an idea science abandoned in the seventeenth century. (See chapter 10.)

For that reason, I hope you will use this book as a gateway to reading more about science. Incredible stories are all around you, and if you scratch a bit, you'll find something fascinating. Take a look at tomorrow's newspaper, or take one of the subjects in this book and cruise the Internet. (For starters, explore my website thatsnotinmysciencebook.com, which contains even more bewildering science facts.) From the new medications being invented to the cholesterol-lowering foods being created to the exploration of the universe that is underway, if you want knock-your-socks-off news, just keep reading the science stories!

In the meantime, forgive me for all the subjects I didn't have room for. If you want more, write the publisher, and maybe there can be a *That's Not in My Science Book Volume II.* Enjoy Volume I! I encourage you to skip around as you read, but give some of the not-quite-your-cup-of-tea topics a try. I promise you'll find them interesting!

PART ONE

Secrets of the Earth Revealed

1

How Old Is the Earth Anyway?

The Geologic Time Scale

One of the facts that is presented very clearly in science books, newspapers, and magazine articles is that the earth is about 4.6 billion years old, give or take a few years. But what they don't talk about is how they *know* the earth is that old. No one was around then, so how did earth's "timekeepers" get from there to here, marking off all the eons, eras, periods, epochs, and ages in between?

The Greeks had their theories (as you will see in this book, not much got past them unnoticed). In the fifth century BC, the Greek historian Herodotus made one of the earliest recorded geologic observations. Finding fossil shells far inland in what is now parts of Egypt and Libya, he correctly inferred that the Mediterranean Sea had once extended across much of the land farther to the south. Unfortunately, this idea caught the interest of few, so no other historian or scientist until the seventeenth century explored how shells came to be found inland, nor did anyone do any conclusive work on how or when the planet was formed.

The Geologic Time Scale

Today scientists understand the earth's age well enough to have created a geologic time scale, which is the standard method used to divide the earth's long natural history into manageable parts. (Even then, however, because the units of time are so long, geologists tend to talk in terms of upper/late and lower/early and middle parts of certain time spans.)

Different spans of time on the scale are usually established

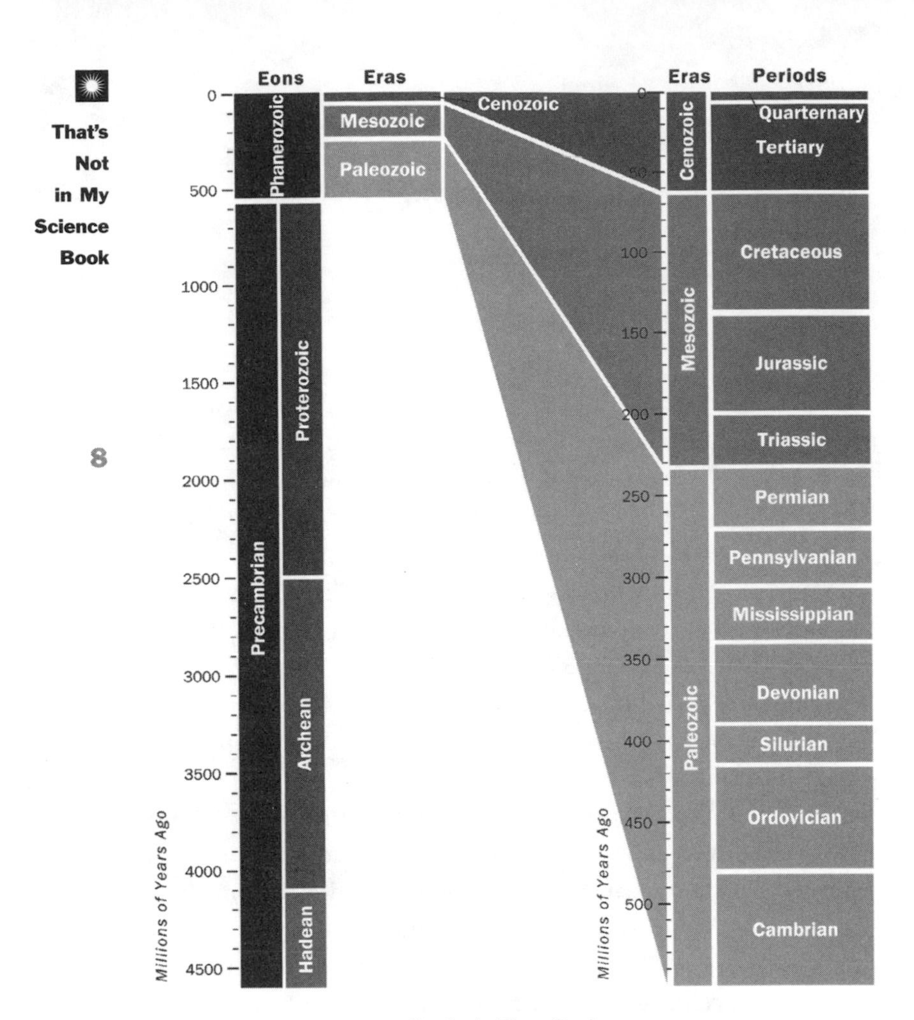

Geologic Time Scale

by major geologic or paleontologic events such as mass extinctions. The first people to measure the earth's age did so using *relative* time—they could tell that some things were older than others. Later, technology allowed scientists to better assess *absolute*, or more exact, time. *Think of relative time as physical subdivisions of the rock found in the earth's stratigraphy, and absolute time as measurement that determines the actual time that has expired, with mass extinctions as the means to separate the spans of time.* As the earth changed, the organisms that could live there also changed.

Now the history of the earth is broken up into a hierarchical set of divisions; in descending order of time length, they are eon, era, period, epoch, and age. The first two eons are the Hadean eon (4.6 to 3.8 billion years ago), with no evidence of life, and the Archean eon (3.8 to 2.5 billion years ago), with signs of bacteria and blue-green algae. Then we enter the "-zoic" eons; the *zoic* part of the word comes from the root *zoo*, which means "animal." The Proterozoic eon (2.5 billion years ago to 542 million years ago) marks the time when protists, algae, and soft-bodied creatures (such as worms and other animals that have no skeletons) began. The Phanerozoic eon (approximately 545 million years ago) was originally thought to be when life began, but later scientists realized that it was when animals evolved to have shells or internal skeletons that permitted the formation of more readily identifiable fossils. The Phanerozoic eon includes three eras that are particularly significant to us:

- The Paleozoic (545–251 million years ago) is the oldest and is when life began to diversify with the development of fish, amphibians, land plants and animals, and basic reptiles.
- The Mesozoic (251–65 million years ago) marks the age of dinosaurs when the major vertebrate life-forms were large reptiles.
- The Cenozoic (65 million years ago to the present) begins with the extinction of the dinosaurs and continues to the present day, when mammals dominate the earth.

How the Relative Time Scale Developed

As in most other scientific developments, the ability to date the earth was discovered not by a single person but by a succession of people who built on what others had learned.

Surprisingly, shark teeth were an early clue to the earth's history. In the seventeenth century, Nicholas Steno (1638–1686), a Danish academic who studied medicine, moved to Florence, Italy, where he gained an official position with the Grand Duke of Tuscany, Ferdinand II. When two fishermen

As this nineteenth-century engraving indicates, early fossil hunters had little concept of geologic time and often assumed that dinosaurs that had in fact lived millions of years apart were from the same era. Source: *The Iconographic Encyclopaedia of Science, Literature, and Art* (1851).

caught a particularly large shark in 1666, Duke Ferdinand ordered that the shark's head be sent to Steno for dissection. Steno's study of the shark's teeth led him to note their resemblance to similar stony objects, "tongue stones," that he had observed in certain rocks. He put forth that the tongue stones in the rock looked like shark teeth because they *were* shark teeth. Yet what were they doing embedded in rocks far from the sea? Though he was not the first to observe that these "rocks" had at one time been living things (what we now know as fossils), this connection led Steno to more closely examine the possibility, and he also began to explore how one solid object such as a tongue stone could come to be found inside another solid object such as a layer of rock.

By 1669 Steno was able to describe two basic geologic principles. The first stated that sedimentary rock is laid down in a horizontal manner (in strata). The second was that younger rock is found on top of older rock, and it appears in that order unless the layers were later disturbed by the formation of a mountain or a cave. This was a major contribution to science, and it became known as *Steno's law of superposition* (now known as the *principle of superposition*).

As you might guess, Steno's discovery was a measurement of *relative time*, not *absolute time*. In principle, two rock layers could form millions of years apart or a few thousand years apart, and in Steno's day, scientists had no way of concluding the exact "when" of what happened.

Classification of Rocks Contributes Another Key

During this time, a very influential teacher of mining and mineralogy, Abraham Gottlob Werner (1749– or 1750–1817), was teaching in Germany. Werner used his background in mining to develop techniques for identifying rocks and minerals. He determined that different types of rock were formed during different periods, and he championed classifying the rocks of the earth in five groups: Primitive (ancient rocks with no fossils that were believed to precede the biblical Flood), Transition (the very first orderly deposits from the ocean), Secondary (rocks containing fossils), Alluvial or Tertiary (sediments believed to be deposited after the Flood), and Volcanic (associated with volcanic vents).

Because Werner was well respected, students came from all over Europe to learn from him. His classification ideas became well known, which was a good thing—scientists were able to continue the classification process, and his efforts laid the groundwork for classifications that are still used today.

However, while Werner's popularity put scientists on the correct path in one area, it permitted him to lead his followers astray in another: Werner put forward a Neptunian view of the earth, claiming that at one time there had been an all-encompassing ocean (a result of the biblical Flood) that had deposited all the rocks and minerals across the earth's crust. So the World According to Werner made him a formidable opponent to the

Plutonists, who felt that volcanoes and earthquakes contributed to the changing face of the planet. Through the end of the eighteenth century and well into the nineteenth century, there was a highly contentious battle among these two opposing forces, and Werner's high standing gave a great deal of credence to what later proved to be an erroneous theory.

Earth's story was still waiting to be told.

Learning More and More

As you may have noticed, the people making the discoveries about the earth came to their work from random backgrounds, and James Hutton (1726–1797) was no exception. Geology was not yet a field of study, and Hutton, who was born in Edinburgh, had trained to be a doctor. However, an inheritance of a small farm in Berwickshire changed the course of his life. As he learned about farming, he became intrigued with the study of the earth's surface, and eventually he returned to Edinburgh to devote his time to mineralogy and tracing the origins of various minerals and rocks.

Drawing upon work done by scientists in Italy, Hutton developed a conception of the dramatic upheaval necessary to form the land as we know it—he felt that present rock layers are formed because of dramatic changes in the earth's surface. Hutton's theories began to combat the Neptunists' belief that all rocks had been distributed by the biblical Flood. Hutton also became convinced that subterranean heat could upend layers of rock, causing the dislocation of strata and the possibility that water or molten rock could travel through veins created during the upheaval to re-form the earth's surface. The most important concept that he championed was that of uniformitarianism, that natural geologic processes are uniform in frequency and magnitude throughout time. In other words, changes that are occurring today are very much like changes that occurred in the past, and therefore, geologic phenomena can be interpreted based on current observations. (This went against the ideas of the catastrophists, who believed that a single cataclysmic event, such as a major worldwide flood, caused the current structure of the planet and that little changed after that.)

Hutton also supported the idea that rocks could be classified according to relative age—that each layer represented a specific interval of geologic time, with the bottom layer being the oldest. While these ideas are obvious to us today, this was revolutionary reasoning two hundred years ago. Of course, people also realized at that time that there was no way to measure the length of time between rock deposits.

Hutton wrote his thoughts in a paper (*Theory of the Earth, or an Investigation of the Laws Observable in the Composition, Dissolution and Restoration of Land upon the Globe*) delivered to the Royal Society of Edinburgh in 1785. Though Hutton's thinking was brilliant, his writing was dense and difficult to understand, so it took the efforts of a friend, John Playfair, a University of Edinburgh professor, to take the information and make it understandable, which he did in a book published five years after Hutton's death: *Illustrations of the Huttonian Theory of the Earth.*

Understanding Rock: The Work of a British Surveyor

William Smith (1769–1839), a British civil engineer and surveyor, was born into a family of farmers in Oxfordshire, England, and is often regarded as one of the greatest of the early geologists because of his careful observation of rock. He apprenticed to be a surveyor, which gave him the opportunity to travel all over England, surveying canal routes. In their work, Smith and other surveyors needed to understand the rocks through which canals were to be dug. This led Smith to observe that fossils found in a section of sedimentary rock were always in a certain order from the bottom to the top of the rock section. This ordered appearance recurred no matter what part of England he visited. As a result, Smith is credited with the *principle of faunal succession*, that strata from various locations can be correlated because the fossils appear in a definite sequence. Smith was the first to create geologic maps using the fossils as a tool for mapping the stratigraphic order of the layers, not their composition.

Geology Is Finally Given Proper Standing

Charles Lyell (1797–1875), a British scientist who is often called the father of modern geology, wrote *Principles of Geology* (published between 1830 and 1833) and showed that the earth has changed slowly and gradually through the ages and is still changing.

Lyell was born in Scotland, and after training to go into the law, he abandoned that field in order to further explore geology. He and John Playfair were major advocates of Hutton's then-controversial idea of uniformitarianism—that the earth was shaped by slow-moving forces that acted on it over a very long period of time.

By the middle of the nineteenth century, Lyell's *Principles of Geology* had become a very influential work, and for his efforts, Lyell was knighted in 1848 and made a baronet in 1864. One of his close friends was Charles Darwin, and though Lyell did not fully embrace Darwin's theory of natural selection, he was supportive of his friend's efforts on *The Origin of Species.* As it happens, Lyell's push for the understanding of how very long the geologic time scale was thought to be (we now know it's even longer than Lyell realized) was absolutely key to permitting serious consideration of Darwin's theory of natural selection. Only with this long time frame could the evolutionary process Darwin described even be considered.

Then Along Came Carbon Dating

Following Lyell's time and for the next one hundred years or so, the age of the earth and the rock strata was the subject of considerable debate. Creationists felt the time span was only a few thousand years; others suggested much larger spans of time. Finally advances in the latter part of the twentieth century allowed radioactive dating—the measure of radioactive decay—to provide relatively firm dates to geologic horizons.

The fellow who came up with the system for radiocarbon

dating was Willard Libby (1908–1980), a native Californian who became a college professor. He was quite fascinated by the study of radioactivity and early on did a great deal of work using Geiger counters.

During World War II, Libby joined the Manhattan Project and was responsible for working with the uranium-235 that was used in the atomic bomb dropped on Hiroshima. After the war, he went back to Berkeley to teach and began studying with radioactive carbon-14 (discovered by another scientist in 1940). By 1947 Libby had observed that plants absorbed some of this trace carbon-14 during photosynthesis. A living plant absorbs the same amount of C-14 in photosynthesis that would naturally spontaneously decay or change into N-14 (nitrogen-14). This means that the amount of C-14 in a living plant remains a constant, and since animals eat plants, they would also have the same constant amount of C-14 in their bodies. (This includes human beings.) However, once the plant died, it couldn't absorb any more carbon, and the carbon-14 it contained would decay at a predictable rate. By finding the concentration of carbon-14 left in the remair [illegible] plant, Libby discovered he could calculate how much tir [illegible] passed since the plant died.

He started by testing his process on objects of known age. He soon found that a reliable age could be found for carbon-based artifacts—wood, parchment, textiles—up to 45,000 years old because carbon-14 disappears atom by atom on a very exact schedule. This has allowed estimates of the age of Egyptian mummies and prehistoric dwellings, among other things. In 1960 he was awarded the Nobel Prize in Chemistry for leading the team that developed carbon-14 dating.

Carbon-14 dating has been instrumental in mapping human history. However, objects more than 45,000–50,000 years old don't have enough carbon-14 to measure, so now scientists have found they can use similar techniques that use radioactive isotopes with a longer half-life than carbon-14. Among the most widely used currently are potassium-40 with a half-life of 1.25 billion years, uranium-238 with a half-life of 4.5 billion years, and rubidium-87 with a half-life of 49 billion years.

So How Do We Know the Age of the Earth?

All of this information about ways to determine the age of the earth brings us back to the original question: How do we know how old the earth is?

At this point, there is no way to pinpoint the age of the earth exactly because it is thought that earth's oldest rocks have been recycled and destroyed by the process of plate tectonics (see chapter 2). However, scientists have been able to determine the probable age of the solar system, and thereby calculate an age for the earth by assuming the earth and the rest of the solid bodies in the solar system were formed at the same time. (It is actually easier to date the moon because it has not been disturbed by plate tectonics, and so more of its ancient rocks are available.) Though scientists have found rocks on earth as old as 4.03 billions years (in northwestern Canada), it is thought that the earth is actually about 4.6 billion years old, which, based on what we know of the universe, would be consistent with current calculations of 11–13 billion years old for the Milky Way and 10–15 billion years for the age of the universe.

But keep following the news for updates. In 1933 people thought the earth was only 2 billion years old. New scientific methods and advances in technology may change this "birthday" calculation yet again. (Any shifts or changes in the time scale are overseen by the International Commission on Stratigraphy.)

And while the geologic time scale is very much still a work in progress, one of the elements scientists have been pleased about is that, since the 1970s, the refinement of the measurements of the ages of things within the Tertiary period (a major division of the Cenozoic era, the most recent of the geologic eras) used to vary from 20 to 30 percent; now it varies by much smaller amounts (5 percent)—meaning that the accuracy of the processes is improving.

2

The Earth Moves in Mysterious Ways

Plate Tectonics, Earthquakes, and More

The satellite photographs of earth taken from space that we have access to today give us a decided advantage in understanding our planet. But as early as 1596 a Dutch mapmaker, Abraham Ortelius, was working with information that was accurate enough to suggest to him that at one time the continents may have "fit together." Ortelius theorized that the Americas were "torn away from Europe and Africa . . . by earthquakes and floods" (from his *Thesaurus Graphicus* online at pubs.usgs.gov/gip/dynamic/historical.html). However, at the time, no one had the additional information necessary to even begin to explain how or whether this could have happened.

It wasn't until more than 350 years later that scientists began to more fully understand why the continents look like one big spread-out jigsaw puzzle. For example, look at how South America could nestle in under part of Africa; and note how Greenland appears to be the connecting piece between Europe and North America.

By the 1960s, an early theory of continental drift was leading to plate tectonics, the theory that the Earth's outermost layer is fragmented into a dozen or more large and small plates that are moving in relation to one another, riding atop hotter, more mobile material. In geologic terms, a plate is a large, rigid slab of solid rock moving on a "plastic" asthenosphere. And the root of *tectonics* comes from the Greek word for "to build." Now virtually all scientists agree that this latest theory takes us a long way toward explaining how the continents started out—and why they are where they are today.

Just as the periodic table provided an organized way of arranging the chemical elements and revolutionized chemistry, and the discovery of the genetic code changed the nature of the study of biology, the theory of plate tectonics, only finalized in the late 1960s, has revolutionized geology.

How It Came to Be

Scientists of the seventeenth and eighteenth centuries addressed many questions, including how the earth's continents came to be. Like the Dutch mapmaker in the late sixteenth century, an eighteenth-century German theologian tried to point out that the coastlines of South America and Africa looked like they had been torn apart.

By 1859, a French scientist, Antonio Snider-Pellegrini, introduced the idea that all the continents were once connected together during the Pennsylvanian period (314–280 million years ago). Scientists who were contemporaries of Snider-Pellegrini were beginning to locate similar plant fossils on various continents, which led them to consider the possibility that one huge landmass might have existed at one point. Snider-Pellegrini's explanation for what happened followed the catastrophist thinking of the day: He felt that the Flood (of biblical fame) had pushed the continents apart.

The "puzzle piece" issue was taken a step further in 1912, when a scientist-explorer named Alfred Wegener (1880–1930) put forth the idea of continental drift. Working as a lecturer in astronomy at the University if Marburg, Wegener came upon a scientific paper that noted the existence of identical plant and animal fossils on two sides of the Atlantic. Wegener began to contemplate this discovery along with the puzzle piece observations that he and others had noted.

Wegener determined that until the Carboniferous (or coal-forming) period about 300 million years ago, the continents were a single supercontinent—he referred to it as Pangaea (from the Greek for "all the earth"). Wegener explained that Pangaea then split for some reason, and its pieces have been moving away from each other ever since. Wegener based his theory on several pieces of evidence including geologic, paleontologic, and climatological factors:

- *The composition of the Mid-Atlantic Ridge that forms such islands as Iceland and the Azores.* (The Mid-Atlantic Ridge is a mostly underwater ridge of mountains that extends the length of the Atlantic Ocean—from the Arctic Ocean in the north to the Antarctic area in the south. At some points the mountains break through the water level to form islands.) It is thought that the Mid-Atlantic Ridge was formed by a divergent boundary that separates tectonic plates. Wegener felt that the Ridge was material left behind when the continents that now flank the Atlantic broke apart. He also noticed how mountain ranges, mineral and rock types, and glacial deposits match up when continents are envisioned together, forming a continental jigsaw puzzle.
- *The surprising distribution of fossil remains of trees and other plants that were around during the Carboniferous period, which had been noticed by scientists before him.* Though each region of the earth seems to have its own specialized vegetation, botanists have found that some plants such as the tropical *Glossopteris* (the seed fern) had thrived in widespread locations, and its fossils are found in India, Australia, South America, and South Africa and also in coal seams in mountains near the South Pole.
- *Animal distribution.* Wegener felt that a land called Lemuria once linked India, Madagascar, and Africa, and this would explain the widespread distribution of the lemur and the hippo. The fact that marsupials such as the kangaroo and the opossum are found only in Australia and the Americas made Wegener also link Australia with distant South America.

Drawing upon his knowledge of various aspects of science as well as his own explorations, he gave a talk on continental displacement to the Frankfurt Geological Association in 1912. By 1915, he was ready to write what would be the first of four versions of *The Origin of Continents and Oceans.* While Wegener continued to come up with evidence that backed up his theory, he was unable to explain what would have pushed the continents apart. He finally concluded that the continents were like great barges, plowing their way to their current positions like icebreakers. Most scientists, however, found this explanation to be preposterous and dismissed the whole theory as a result.

In the autumn of 1930, Wegener, who was also trained as a meteorologist, agreed to accompany a scientist friend to help establish a weather station in Greenland. Though Wegener reached the intended destination despite horrific weather conditions, he died—it is thought he had a heart attack—when he left the station to go for more supplies. Otherwise, he might have lived long enough to see that scientists eventually embraced his theory.

Just prior to Wegener's death, British geologist Arthur Holmes became an active supporter of Wegener's theory and proposed that over a long period of time the heating and cooling of the earth (thermal convection) might be enough to break apart land masses and cause the continents to move. But nothing about this idea was catching on. Holmes, too, was totally ignored until the 1960s.

The Story Was Shelved for Several Decades

In the 1920s Wegener had predicted that many mysteries would remain until technology was created that would permit study of the ocean floor, and his prophesy was correct—the ocean floor held the key.

In the 1940s and early 1950s, geophysicist and oceanographer Maurice Ewing began taking seismic readings of the ocean floor (a method for measuring the earth's vibrations). Ewing's readings and measurements were sent back to his research assistant Marie Tharp (1920–), who had been permitted to train as a geologist only because of the shortage of male workers caused by World War II. Working along with colleague Bruce Heezen, Tharp mapped the ocean floor in detail. Their work eventually uncovered a 40,000-mile underwater ridge that encircles the globe. This geophysical data laid the foundation for the conclusion that the sea floor spreads from central ridges and that the continents are in motion, which paved the way for acceptance of the theories of plate tectonics and continental drift.

Today Marie Tharp is considered a pioneer of modern ocean floor cartography, but it has only been since the mid-1990s that she has been recognized for her work. (She was not in anyone's science book, even though she deserved to be.)

In the 1950s scientists also started experimenting with magnetometers (adapted from airborne devices developed during World War II to detect submarines), and they began to notice something unexpected: The magnetic fields on the ocean floor changed directions periodically. Initially, scientists felt this was because basalt, an iron-rich volcanic rock that makes up the ocean floor, contains magnetite that can distort compass readings. However, based upon the study of lava formations in Hawaii, scientists began to realize that as newly formed rock cools, the magnetic materials record the earth's magnetic field at the time. They saw that earth's magnetic field dramatically reverses at intervals (layer by layer) that range from tens of thousands to many millions of years, with an average interval of approximately 250,000 years. The last such event occurred some 780,000 years ago.

This new information gave scientists two new clues about the earth's past history:

1. When molten rock rises from below the earth's crust (in this case, underwater) and hardens into new crust, the iron in the rock hardens with the magnetic pull of the current magnetic pole. As new crust is pushed up at the crest of the ridge, it causes the previous crust to separate, so young rock is near the center of the ridge, and the farther away you move from that ridge, the older the rocks are. (If molten iron is allowed to cool, the molecules of iron line up with the magnetic field of the earth, and the iron forms a natural magnet.)
2. This information combined with the magnetic "striping" on the seafloor provided scientists with one more way to understand how the earth was formed. Symmetrical stripes of rock parallel to the ridge crest alternate in magnetic po-

larity (current direction–reversed–current direction, etc.), suggesting that the earth's magnetic field has flip-flopped many times.

Working with the information that had been acquired by 1962, American geologist Harry Hess proposed the theory of seafloor spreading. If the earth's crust was expanding along the oceanic ridges, then Hess concluded that it must be shrinking elsewhere—eventually (millions of years later) descending into oceanic trenches. Hess suggested that the continents did not float about, but interacted with the oceanic crust. Plate interactions formed mountain ranges, earthquakes, and volcanoes. He also proposed a mechanism for driving the movement of plates.

By explaining both the symmetrical zebralike magnetic striping and the age of the rocks increasing as you travel from the mid-ocean ridge to the continents, the seafloor-spreading hypothesis quickly gained converts and represented another major advance in the development of the plate tectonic theory.

How It Works

Today we know that the earth's surface is made up of eight to twelve big plates and twenty or so smaller ones. They all move in different directions at different speeds, and some are slow moving (a fraction of an inch per year), while others are relatively speedy (a few inches per year) in their progress. You cannot necessarily figure out the plates by knowing the continents. For example, the North American plate roughly traces the outline of the western coast (where everyone knows there is a great deal of earthquake activity), but the eastern part of the plate extends halfway across the Atlantic to the mid-ocean region.

These are main features of plate tectonics:

- The earth's surface is covered by a series of crustal plates.
- The ocean floors are continually moving. New ocean floor is being created, which causes spreading from one

"Bumper" Plates!

Over time, scientists began to realize that the crust of the earth consists of two major layers. The outer shell, the lithosphere, is the crust and rigid mantle, broken into fragments or plates. It floats on a very slow-flowing "plastic" inner shell called the asthenosphere. The movements of the plates reshape continents, build mountains and valleys, and affect the dominance and evolution of species. Because the sides of a plate are being either created or destroyed, its size and shape are continually changing. There are three types of boundaries that define the way the plates bump against each other:

***Convergent boundaries*: These consist of two plates colliding or pushing against one another. The results of these collisions are a bit like the game rock, scissors, paper, in that there is a predictable pattern of dominance. If a very dense oceanic plate encounters a less-dense continental one, the oceanic plate is generally forced underneath, forming a subduction zone. These encounters create changes in the geology of the area that can result in the creation of mountain ranges and volcanoes. The mountainous spine of South America and the Cascade Mountains of North America are good examples of this.**

***Divergent boundaries*: A divergent boundary consists of two plates moving away from one another. Over time, the space is filled with new crustal material from molten magma. An example of this is the oceanic ridge systems that cause ocean spreading, including the Mid-Atlantic Ridge.**

***Transform boundaries*: These occur where two plates slide past one another. These are known as *strike-slip faults*. Friction between the plates stops plate movement. When the plates are pushed and they can't move because of friction between them, stress builds up. When stress reaches a level that exceeds the friction point, there is sudden motion along the fault, causing an earthquake. For example, scientists know that the Pacific plate is moving north while the North American plate is moving south, and this causes friction along the San Andreas Fault along the Pacific Coast. When the plates slip, we get an earthquake.**

---

area (usually the center) and sinking at another area (usually the edges), where the crust is being destroyed.

- Convection currents in the magma (the molten rock material) beneath the crustal plates move them in different directions.
- The source of heat driving the convection currents is radioactivity deep in the earth's mantle.

The Astounding Power of Plate Tectonics

While movement of a fraction of an inch to a few inches per year doesn't seem like much, we only need to remember back to December 26, 2004, to know that earth's movements can result in very dramatic events. On that date, of course, a tsunami hit Asia, and it was what we can only call "biblical" in its power and its devastation, rising up to a terrifying 100 feet in some locations. According to scientists, it was the result of an earthquake with a magnitude of 9.15 that lasted nearly ten minutes. (Most last no longer than a few seconds.) The earthquake and resulting tsunami killed more than 283,100 people, making it one of the deadliest in modern history.

The earthquake occurred in the Indian Ocean, off the western cost of northern Sumatra, Indonesia. The resulting tsunami devastated the shores of Indonesia, Sri Lanka, South India, and Thailand and caused serious damage and death as far away as the east coast of Africa.

According to data from the Lamont Doherty Observatory in Palisades, New York, "an estimated 1200 km (750 miles) of faultline slipped about 15 m (50 feet) along the subduction zone where the India Plate dives under the Burma Plate" (Lamont Doherty Observatory, Earth Institute at Columbia University). Seismographic data indicates that the slip took place in two phases over several minutes and caused a 400 kilometer (250 mile) long and 100 kilometer (60 mile) wide rupture, the longest ever known to have been caused by an earthquake.

The tsunami occurred because the sudden vertical rise of the seabed by several meters during the earthquake displaced colossal volumes of water, resulting in massive flooding and destruction.

Back in the US of A

Normally when talk turns to earthquakes, we quite naturally think of California, the scene of the famous 1906 San Francisco earthquake and the state where we know the San Andreas Fault lies.

What we tend to ignore are the facts. There are forty-one

Houses damaged in the San Francisco earthquake of 1906, shown here in a stereograph. Source: Library of Congress.

states and territories in the United States at moderate to high risk for earthquakes, and no region of the country is immune:

- Alaska experiences the greatest number of the large earthquakes—most of them located in uninhabited areas. One of the largest was near Anchorage in 1964 and measured 9.2 on the Richter scale. In some places the ground lifted 30 feet, and the earthquake set off a tsunami that killed 122.
- The largest earthquakes felt in the United States were along the New Madrid Fault in Missouri, where a three-month-long series of quakes from 1811 to 1812 included three quakes larger than a magnitude of 8 on the Richter scale. These earthquakes were felt over the entire eastern United States with Missouri, Tennessee, Kentucky, Indiana, Illinois, Ohio, Alabama, Arkansas, and Louisiana experiencing the strongest ground shaking.

That said, part of our thinking about California is correct; the state does experience the most frequent damaging earthquakes. The configuration of the state's San Andreas Fault is almost an identical twin to that of the North Anatolian fault that produced the magnitude 7.4 earthquake near Ismit, Turkey, in 1999, which killed more than 15,000, with uncounted numbers buried in the rubble. In addition, scientists have confirmed that downtown L.A. is situated on what is known as a blind thrust fault, a type of fault capable of producing a devastating earthquake. (In a quake caused by a thrust fault, the blocks of earth move diagonally, almost vertically. In a strike-slip fault like the San Andreas, the opposing earth plates slide past each other horizontally.)

The destruction caused by an earthquake depends on the type of ground that is affected by the quake. Earthquakes have been known to liquefy loose rock material that has heavy buildings on it. In a recent California earthquake, the most damage occurred to areas that were built on landfill.

As we learn through any news coverage of major earthquakes, a strong one can collapse buildings and bridges, disrupt gas, electric, and phone service, and sometimes trigger landslides, avalanches, flash floods, fires, and huge destructive ocean waves (tsunamis).

Often, however, the major devastation from an earthquake occurs from the events it spawns: The 1906 earthquake in San Francisco was serious, but it was the devastation of three days of fires that caused the most serious damage. Because of broken water mains it was impossible to fight the fires. Twenty-eight thousand buildings were destroyed; 300,000 people were left homeless, and the death toll was approximately 700.

While the shaking of the earth is frightening, it is seldom the direct cause of death or injury. Collapsing buildings, flying glass, and falling objects are generally the cause of earthquake-related injuries. Knowing this is extremely instructive for making adequate preparations. You will likely be able to withstand the earth tremor; what you need to focus on is creating an environment that diminishes hazards from other risk factors such as falling objects.

As with all natural disasters, the seriousness of an earthquake depends as much on the size of the population of the region as it does on the magnitude of the actual disaster. An earthquake that is 8 or 9 on the Richter scale that occurs in a deserted area would not be as serious as one that measures a 5 or 6 but occurs in a major city. Time of day also makes a big difference. If most people are at home and asleep, fewer people will be hurt than if it's midday and people are in office buildings, on freeways, and otherwise moving about outdoors.

What we do know is that earthquakes are going to continue to shake our world, and the only questions are where and when.

How Earthquakes Are Measured

Seismographs located all over the world measure the shaking of the earth. The measure of earthquakes we all hear about is the Richter scale. In 1935 Dr. Charles Richter, a geologist at the California Institute of Technology, proposed using his method, which is actually a mathematical scale measuring the magnitude of ground movement. Like the waves created by a pebble tossed into water, the ripples of an earthquake weaken as they get farther and farther from the epicenter of the quake. Because an earthquake will affect different areas slightly differently, and it's impossible to measure each spot where an earthquake is felt, the measurement of any earthquake is usually gathered from at least two different seismographs, and a range of the magnitude is presented, such as from a low of 7.6 to a high of 8.5.

About Tsunamis

Tsunamis are ocean waves produced by earthquakes, impact events, landslides, or volcanic eruptions. When a tsunami hits shore, it most frequently does so as a rapidly rising turbulent surge of water choked with debris. The net result is a sudden flood that comes from the ocean.

Tsunamis may be locally generated, or they may come from a distance, and their level of destruction is equally varied. In 1957 a distant-source tsunami generated by an earthquake in the Aleutian Islands in Alaska struck Hawaii 2,100 miles away. Hawaii experienced $5 million in damages. In contrast, in 1992 the Cape Mendocino, California, earthquake produced a tsunami that reached Eureka in about twenty minutes and Crescent City in fifty minutes. This tsunami had a wave height of only about 1 foot and was not destructive.

A tsunami is often incorrectly referred to as a tidal wave, but it is not a surface wave—it is actually a deep ocean wave where the whole ocean area moves vertically several inches or feet. It can travel at speeds averaging 450 and up to 600 miles per hour in the open ocean.

Ironically, tsunamis are not felt in ships because the wave length is hundreds of miles long with an amplitude (height) of only a few feet. However, as a tsunami approaches land and the speed decreases, the amplitude increases. While waves as high as 100 feet have been recorded (such as the 2004 Asian tsunami), generally the waves are 10 to 20 feet high and can still be very destructive.

Areas most likely to experience a tsunami are those that are less than 25 feet above sea level and within 1 mile of the shoreline. To detect a tsunami during its early stage, tsunami detectors are placed near the ocean bottom and are effective if there is a way to spread the word. Though geologists recognized the possibility of a tsunami when the earthquake occurred in December 2004, there was no organized method for getting the word out. A warning system in the Pacific has saved hundreds of lives over time.

The biggest risk with a tsunami are all the risks associated with flooding—contamination of water, property damage, fires from ruptured gas lines, and so on—combined with the threat of drowning because of the very suddenness of it. The lack of predictability as to where and when a tsunami will hit is also problematic. One community may experience no damaging waves while another one an ocean away can be devastated by it.

More Study Under Way

Whether the natural disaster is an earthquake or a tsunami, the first step in survival is anticipation. Even with the overwhelming power of the Asian tsunami of 2004, the communities where someone recognized what was happening and actually responded when they saw how the water had receded so dramatically fared better than those immobilized by ignorance. The same is true with earthquakes.

Science's goal, of course, is to become better at predicting these natural disasters, and technology is paving the way. Today satellite imagery permits greater study of plate tectonics, which should lead to a better understand of earthquakes. And in related work, American astronauts placed several laser reflectors on the moon, which will permit further study of the movements of the plates.

3

Dinosaurs

From Mythical Griffin to Modern-Day Bird

Children find dinosaurs endlessly fascinating, so from early on, children have a chance to become acquainted with these amazing prehistoric mammals—from Barney and friends (light on accuracy, big on association with dinosaurs) to wonderful class field trips to natural history museums. After the elementary school years, however, most school districts have little time in the curriculum to focus on dinosaurs. As a result, children learn about dinosaurs at an age when the study is barely more than a story.

So when it comes to dinosaurs and what isn't in the science book, almost anything goes. This chapter tells you about the early dinosaur finds—an unfortunate but interesting story—and then looks at what scientists today are discovering. Over the past thirty years, a wealth of new information has been uncovered, and the idea of dinosaurs as slow, clumsy beasts has been replaced with a great deal of new information about these reptiles.

It is unfortunate that students don't have more time to discuss dinosaurs—the study of these remarkable creatures is important both for understanding of the causes of past major extinctions of land animals and for appreciating changes in biological diversity.

Early Thoughts on Dinosaur Fossils

For years, people were finding—perhaps tripping over—dinosaur fossils without having any idea what they were. It is commonly thought that many of the legends about monsters,

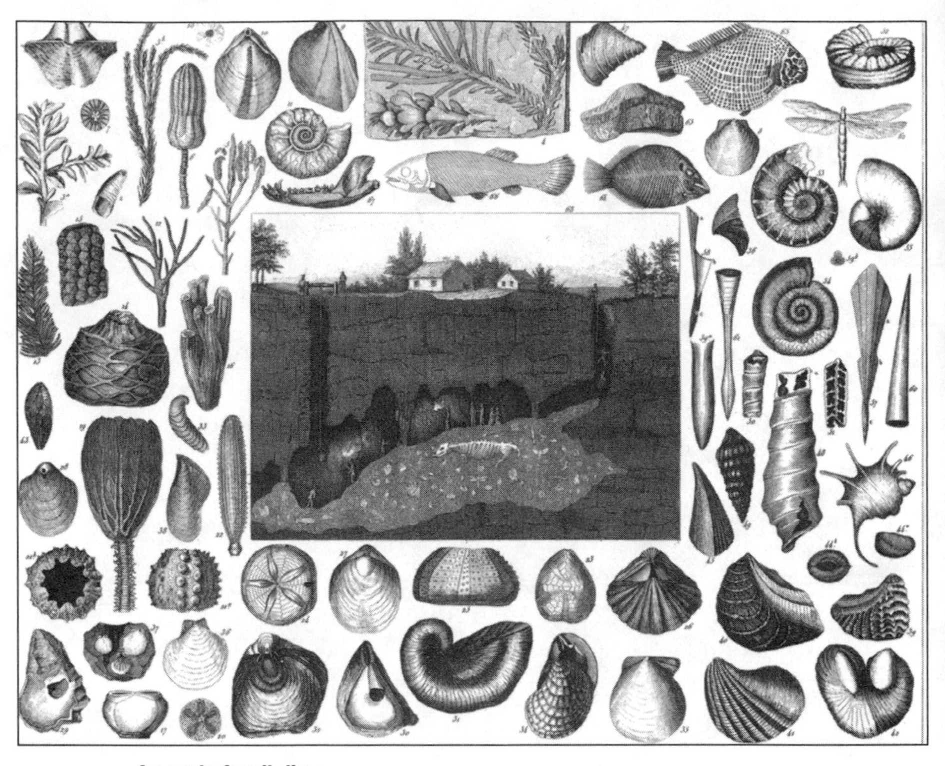

An early fossil dig.
Source: *The Iconographic Encyclopaedia of Science, Literature, and Art* (1851).

giants, and griffins were spawned by these amazingly large bones that people came upon but couldn't explain.

The first documented description of a dinosaur fossil was written in 1676 by Robert Plot (1640–1696), an antiquarian and county historian who wrote and published on the subject of the natural history of his community, Oxfordshire, England. Because no one had any conception of dinosaurs, Plot made an educated guess and identified the bone specimen as part of a leg bone of one of the war elephants that the Roman general Plautius was thought to have brought with him when he invaded Britain in 72 AD. In 1677 Plot revised his thinking to the bizarre idea that the bone was the leg bone of a giant. The bone was eventually lost, but Plot's written description of it—it weighed almost 20 pounds and measured nearly 2 feet around—along with an accompanying engraving has permitted modern-day scientists to identify it as a dinosaur femur, probably that of a *Megalosaurus*.

Another early dinosaur hunter must certainly be mentioned here because for years she was forgotten or ignored because of her gender. Mary Anning (1799–1847) lived in Lyme Regis, along the southern coast of England, and learned fossil hunting from her father, a cabinet maker. He died when Anning was only eleven, leaving the family destitute.

The Lyme Regis area was a rich one for fossil finding. As the sea and the winds tore away at the cliffs, fossils became visible to those who walked along the shore, though that didn't mean the fossils were easy to retrieve. Fossil finds attracted tourism, however, and by her teen years, Mary Anning and her family had built a tidy business selling the fossils they found each day to tourists. Anning walked and waded under the unstable cliffs looking for specimens she could dislodge from the rocks, and she became known for her skill at freeing the fossils carefully and completely.

Over time, Anning built a following of institutions and sophisticated private collectors, but today, it is difficult to fully attribute to her everything that she found. Museums of the day tended to credit those who donated the fossils, so often the name associated with a particular find is the collector, not the person who actually located the fossil. Even with this in mind, Mary Anning is known to have made some amazing finds including a small *Icthyosaurus* (1821) and the first nearly complete example of *Pleiosaurus* (1823) as well as the first British *Pterodactylus macronyx*, a flying reptile. As time went on, Anning won the respect of contemporary scientists, and toward the end of her life she received small stipends from several professional societies. Her obituary (1847) was published in the *Quarterly Journal of the Geological Society*, an organization that would not admit women until 1904.

Interestingly, finds are still being made in the Lyme Regis area because of the continuing erosion of the cliffs.

How Dinosaurs Got Their Name

The term *dinosauria* (roughly translated as "terrible lizard") came from the Greek *deinos*, meaning "fearfully great," and *sauros*, meaning "lizard." It was first used by Sir Richard Owen (1804–1892), a British comparative anatomist. Owen studied a group of fossils and noted the columnlike legs (as opposed to sprawling legs of other reptiles) and the five vertebrae that were fused to the pelvic girdle. In 1842 Owen presented them as a separate taxonomic group, identifying them as a suborder of large, extinct reptiles. No one had yet come up with the thought that they were prehistoric.

Owen intended his information to refute the theory of evolution. (Darwin had not yet written *The Origin of Species*, but evolution had been under discussion earlier.) Ironically, Owen's work eventually helped support the arguments of those who supported natural selection and evolution.

For a time he was prosector for the London Zoo, which meant that he dissected and preserved any animals that died in captivity. This gave him vast knowledge of the anatomy of all types of animals. Owen also became well known for his description of the anatomy of a newly discovered (1847) species of ape—the gorilla. In his description he adamantly rejected the idea that apes and humans could in any way be related.

American Finds

In America, the first known find of a dinosaur fossil was in 1787, when a Dr. Caspar Wistar found a thigh bone in Gloucester County, New Jersey, which since that time has been lost. Just a few years later—in 1800—Pliny Moody found 1-foot-long fossilized footprints on his farm in Massachusetts. Scholars from Harvard and Yale came to see them, and they knew *exactly* what they were: They identified them as footprints that were made by Noah's raven!

But in 1838 an amazing process began: A nearly complete dinosaur skeleton was discovered by workers in a marl pit (a deposit of a crumbly type of soil) on a farm in Haddonfield, New Jersey. Not much happened with the remains until 1858, when a fossil hobbyist named William Parker Foulke was visiting the area and realized the significance of the find. What had not been disturbed by the workers was more fully excavated, revealing an animal larger than an elephant with structural features of both a lizard and a bird. Foulke invited anatomist and Philadelphia museum curator Dr. Joseph Leidy to view the find, and Foulke eventually arranged for the bones to be transported to the Philadelphia Academy so that Leidy could study them. Leidy went on to become a pioneer in documenting dinosaur anatomy, and the Haddonfield *Hadrosaurus* is still on display at the Philadelphia Academy of Natural Sciences. That find was to mark the beginning of paleontology—the study of the forms of life that existed in prehistoric times.

The Bone Wars

Paleontology began with a rivalry between two men, Othniel Marsh and Edward Cope, who vied to become recognized as the ultimate authority in this new field. This feud may have hastened the finding of dinosaur bones in North America, but because the decades-long battle involved lying, skullduggery, politics, and cheating, it was certainly a regrettable example of how to get something done.

Two wealthy and competitive American scientists, Othniel Charles Marsh (1831–1894) and Edward Drinker Cope (1840–1897) began their relationship as friends. They had met at the University of Berlin and discovered they had a common interest in fossils. When they returned to the United States, Marsh became a professor at Yale, where he encouraged his uncle, George Foster Peabody, to support his explorations. (The Peabody Museum in New Haven is the repository of many of Marsh's finds and is still a well-respected natural history museum.)

Cope got a job working for the preeminent paleontologist Dr. Joseph Leidy, who was at the time busy analyzing the Haddonfield find. As he got involved in the project, Cope became so excited about the New Jersey discovery that he moved his family to Haddonfield so that he would be able to work directly in the pits.

In 1866, Marsh, who by this time had been appointed America's first college professor of paleontology, came to Haddonfield to tour the pits with Cope. However, Cope later learned that Marsh sneaked back to the site on his own and offered bribes to Cope's workers to send Marsh the finds instead. The result was a bitter feud, known as the Bone Wars—spying, bidding up the price of bones, attacking the validity of the other's work—in short, doing all that they could to destroy each other.

During the next twenty years, Cope and Marsh moved on from New Jersey to launch separate expeditions to the West, using U.S. cavalry forts as staging areas and mule-drawn covered wagons as vehicles. On Marsh's first expedition to Wyoming in 1870, William F. Cody (of Buffalo Bill Show fame) acted as a guide for the first part of the journey. (He remained Marsh's friend, visiting Marsh whenever his show played in the Connecticut area.) Little regard was given to Indian rights.

In 1879 when railroad workers contacted Marsh about a big fossil find near Como Bluff, Wyoming, Cope arrived, too, and accused Marsh of stealing his fossils. Marsh directed that the dinosaur pits be dynamited rather than allow fossils to fall into the "wrong" hands.

All in all, the rivalry resulted in depleting both men's fortunes. Cope had to sell off part of his collection, and Marsh had to mortgage his house and beg Yale for a salary since he had totally run through his uncle's endowment.

At the beginning of the Bone Wars, there were only 9 named species of dinosaur in North America; by the time these two men died, the number of species was decidedly greater. Marsh discovered a total of 86 new species, while Cope discovered 56. Together they uncovered 136 new species (there was some overlap), creating a new field of science and great excitement the world over.

The Last of the Great Dino Hunters

Named after circus impresario P. T. Barnum, Barnum Brown (1873–1963) spent sixty-six years working in paleontology and was responsible for much of the collection at the American Museum of Natural History in New York. Though Brown built a large collection of dinosaur bones over his long career, he will always be held dear by the public for finding the first *Tyrannosaurus rex*, "King of the tyrant lizards," in Hell Creek, Montana. In 1902 he found the first skeleton, and in 1908, he located one that was better preserved.

An interesting footnote to history: During World War II, Americans became concerned that if the Germans were to bomb New York, the *T. rex* fossils might be lost, so the bones from the 1902 discovery were shipped to the Carnegie Museum in Pittsburgh for safekeeping.

New Theories about an Old Species

Are modern-day birds actually descended from dinosaurs? A growing number of scientists think so.

In the late 1860s, British naturalist Thomas Henry Huxley examined an exciting new bird fossil called *Archaeopteryx* and noticed a skeleton bone that resembled a bone found in theropods, the family of predatory dinosaurs such as *T. rex*. This observation led to his theory that birds evolved from dinosaurs. (We now know that theropods share over 50 anatomical features with modern birds including a wishbone, swiveling wrists, and three forward-pointing toes.)

The bird-dinosaur link was widely ridiculed until a few years ago when Canadian paleontologist Philip Currie and Chinese geologist Ji Qiang published papers on the discovery in China of two small dinosaur fossils that appeared to have feathers. In 1996 a chicken-sized dinosaur known as *Sinosauropteryx* and in 1997 a roadrunner-like dinosaur, known as *Caudipteryx*, appear to provide additional links.

Scientists who support this theory now suggest that as many as half of all dinosaur species during the Cretaceous period (65–144 million years ago) may have had feathers of some sort. Paleontologist Philip Currie believes that even *Tyrannosaurus rex* may have had feathers. Did they fly? Skeletal evidence suggests that these feathered dinosaurs lived solely on the ground, so the feathers may have been for insulation, for display, and possibly with some creatures for some form of very rudimentary flight.

Another key piece of evidence linking birds to dinosaurs came from Madagascar in 1998. Catherine Forster, a paleontologist from the State University of New York, Stony Brook, found a fossil bird dating back 65–70 million years. The bird had a sickle-shaped claw at the end of a thick second toe on its hind feet, and the only other known animal with such "killing claws" are dinosaurs such as velociraptors—these were the dinosaurs in the movie *Jurassic Park* who used their slashing claws during attacks.

Most scientific theories have their dissenters, and the theory of the bird-dinosaur link is no exception. Dr. Alan Feduccia, an ornithologist and evolutionary biologist at the University of North Carolina, says that he and most ornithologists feel that birds and dinosaurs have a common reptilian ancestor, but he feels that birds are not, in effect, living dinosaurs.

Stay tuned and keep reading the news!

An Accidental Piece of Great Luck

An amazing find from 2003 has recently been announced. That year a *T. rex* was found by a fossil-hunting team led by John R. Horner, a paleontologist with the Museum of the Rockies at Montana State University. The *T. rex* skeleton was excavated from the Hell Creek Formation, in sandstone laid down about 70 million years ago, in a remote corner of the Charles M. Russell National Wildlife Refuge in Montana.

Because of the rough terrain where the fossils were found, the only way to get the heavy rock-encrusted bones out for further study was by air. Though this particular tyrannosaurus was estimated to be only eighteen years old at death and was smaller than many, the rock-encased leg bones were too long to travel by helicopter. Scientists decided that they had to carefully break the longer bones to transport the skeleton back to the laboratory. When the broken thigh bones arrived at the laboratory in Bozeman, Montana, luck continued to be on science's side: No one applied preserving chemicals right away, as would usually be done. Though that would have preserved the specimen, it would have contaminated the tissue they soon were to discover. As it happened, the scientists that day began studying the fragments of the bone, only to discover that there were still the remains of soft tissue fragments (bone cells and blood vessels) lining the bone marrow cavity. This type of tissue had never before been recovered from a dinosaur. The examining scientist, Dr. Mary Schweitzer, noted that what they had, amazingly, was the remnants of soft tissue from a prehistoric creature that lived 68 million years ago.

Early studies with a scanning electron microscope showed the dinosaur's blood vessels to be very similar to those of flightless birds like ostriches or emus, which buttresses the theory of birds being living descendants of some dinosaurs. After careful study, the discovery team has concluded that the estrogen-derived tissue was similar to substances now present only in living birds. Continued study of this *T. rex* led to a later announcement that she had been ovulating.

More Discoveries about Ancient Beings

By the 1900s major natural history museums were putting together their own dig teams and continued to identify different species. However, in recent decades, the research has shifted from finding and classifying to studying the lives and habitats of the dinosaurs. In the late 1960s, Robert Bakker proposed that these ancient creatures may well have been as agile and energetic as warm-blooded animals, and the discoveries have proceeded from there.

New discoveries have also come about because of new technology. The use of digital X-rays and CT scans on dinosaur bones has led to a greater focus on the lives of the dinosaurs, and computer technology has permitted scientists to compute various dinosaurs' gaits and speeds.

Among some of the new findings are these:

- One of the richest new fields of exploration has been the Gobi Desert in southern Mongolia, which has been more carefully explored since Mongolia declared its independence from the Soviet Union in the early 1990s. After three disappointing visits, U.S. paleontologists came upon one area where they found sixty dinosaurs in three hours.
- Dinosaur embryos 190 million years old found in South Africa include a *Massospondylus* still curled up inside a shell that was less than 3 inches long. New research suggests that hatchlings began life moving on four legs but somehow matured to walk on two legs, a pattern of development that was almost unheard of among vertebrates, say researchers headed by Robert Reisz of the University of

Toronto. The fossils were actually excavated in 1978 in South Africa, but it took this long to expose the embryos from the surrounding rock and eggshell and analyze the remains.

What Happened to the Dinosaurs

Evidence indicates that we have had at least five mass extinctions. These extinctions occur when there is a change in the geology or climate of an area. Eventually, new better-adapted organisms replace the organisms that were there. These extinctions usually occurred quickly in geologic time, which for us would be around one million years or so. Each group of "new" organisms includes those with traits that are suited to the new environment, which is a result of the change in climate or geology that caused the latest extinction.

- 544 million years ago soft-bodied organisms became extinct.
- 443 million years ago small marine organisms were wiped out.
- 362 million years ago many tropical marine species became extinct.
- 251 million years ago the largest mass extinction event in earth's history occurred, affecting a number of species, including many vertebrates.
- 65 million years ago the mass extinction we think of as having brought an end to the age of the dinosaurs occurred. Though mammals and plants were hit hard, they were not wiped out. Placental mammals appeared.

Today it is commonly felt that a meteorite did slam into the earth about 65 million years ago, setting off the events that led to the extinction of the dinosaurs. However, scientists also feel that it is a more complicated story than "lights out" and mass extinction.

Estimates are that a large object, probably an asteroid that may have been 5 or 6 miles across, blasted a 120-mile crater at the tip of what today is Mexico's Yucatan peninsula. (While the magnitude of this seems impossible, recent examples of impacts to other planets have given credibility to this theory. In 1994, Jupiter was struck by a series of cometary fragments, and some of the blasts were larger than the earth's diameter.) This incredibly powerful impact kicked up a worldwide, sunlight-blocking shroud of dust that over time wiped out some 70 percent of the existing plant and animal species—including dinosaurs. However, other factors such as the release of volcanic gases, climatic cooling, sea level change, low reproduction rates, poison gases from a comet, or changes in the earth's orbit or magnetic field may have contributed to the extinction.

4

The Elephant

From Hairy Mammoth to Great Communicator

One of the problems with studying science is that there is so much material to cover there is no time left to contemplate the truly wonderful world around us. Take, for example, the elephant. We may have seen one perform some balancing tricks at the circus, and we are likely familiar with them from the zoo, but neither of these experiences does justice to these largest of land mammals, a species that has a well-defined social structure and strong family bonds. They are fascinating animals to study because they have a complex communication system, a full range of emotions, and bodies perfectly designed for their lifestyle.

Why We Need Elephants

Elephants require broad territories to roam. They need to have plenty of food available, and they need to travel far enough to meet up with other elephant populations so that they don't become inbred. While elephant needs are not simple, there is no end to the good that they do. If given enough space to live, elephants are enormously helpful to those around them and play a vital role in the ecosystem. Neighboring animals drink from the waterholes that elephants dig in dry areas, and when elephants pull down branches, they put more food within reach of smaller animals. Elephants eat branches as well as leaves; this helps prune bushes and stimulates growth. The pathways elephants clear in savannahs and forests provide firebreaks to help slow the spread of wildfires and conduits for rainwater, and the pathways are used by other animals as well as future generations of elephants. Since elephants do not digest all that they eat, some plant matter, including seeds, goes directly through their systems. As they travel and defecate, they spread seeds. Unfortunately, if elephants cannot find enough to eat, they may ravage farmlands, and they can turn areas into dust bowls.

Sadly, our exploding world population has encroached on the elephant territories, and the killing of elephants for their tusks has greatly reduced their population.

About the Elephant

Elephants are the largest living land mammals. Commonly, elephants have been described as either African or Asian (formerly known as Indian). The two species are assumed to be similar, but they are actually quite different genetically. Asian elephants are more closely related to the extinct mammoth than to the African elephant. The African elephant, too, is not what has been presumed: There are actually two separate species: the savannah (or bush) elephant and the forest elephant, which differ from each other primarily in size. The average adult savannah elephant is over 8 feet at the shoulder, and the forest elephant is smaller.

The Asian elephant and the two African species also differ in both body and ear size. The Asian, *Elephas maximus*, has smooth skin, an enormous domed head with relatively small ears, and a sloping back. The African species, *Loxodonta africana*, has very wrinkled skin, a straighter back with a slight dip in the middle, enormous ears, and one more vertebra in the lumbar section of the spine.

Their trunks, too, are different. The African elephant has two "fingers" at the end of the trunk, making it easier to pick up items. The Asian elephant has only one "finger." In addition, all adult African elephants have large tusks, while only adult Asian males do. (As explained later, the tusks have led to great problems for the elephant.)

All elephants have distinctive ear patterns that the human eye can recognize as different. Scientists often document studies of various elephants based on their ears.

Source: *The Iconographic Encyclopaedia of Science, Literature, and Art* (1851).

African elephants live in the grassland savannah and open woodland areas of central and southern Africa. The Asian elephants live in India, Nepal, and Southeast Asia, in the scrub forest and rain forest edge.

The elephants' swaying style of walk comes because they move both their front and back legs on the same side forward at the same time. Their feet are well padded and flexible, and an elephant can walk relatively silently because the weight is so well distributed.

Elephants eat all types of vegetation, from grass and fruit to leaves and bark and roots, and they need to consume about 30 gallons of water and 220 to 440 pounds of food per day. Like the panda, they spend almost all day—eighteen hours—feeding themselves. *Pachyderm*, a word used for elephants (and hippopotamuses, too) means "thick skin," and the description is entirely appropriate. Elephant skin is actually a full inch thick over some parts of its body. Surprisingly, their skin is also very sensitive—they spray themselves with water and take dust and mud baths in an effort to protect their skin from sun and biting insects.

Elephant Facts

SIZE: African elephants are larger than Asian elephants. African elephant females can weigh up to about 8,000 pounds (3,600 kilograms) and males up to about 15,000 pounds (6,800 kilograms), though the largest elephant on record was an adult male that weighed 24,000 pounds (10,886 kilograms). ***A full-grown African elephant can be taller than a school bus and weigh as much as four cars.***

Among Asian elephants females are about half the size of males. Females can weight up to 6,000 pounds (2,720 kilograms), and males can weight up to 11,000 pounds (4,990 kilograms).

LIFE SPAN: African elephants can live up to 50–70 years; Asian elephants tend to live closer to 40–50 years.

SPEED: An elephant normally covers about 2–5 miles per hour (3–6 kilometers per hour) and can charge at 25 miles per hour. In comparison, a zebra can reach speeds of 40 miles per hour.

EYELASH LENGTH: As long as a human hand!

An Elephant's Trunk

The elephant trunk is as essential to an elephant as hands are to a human. In addition to smelling and breathing, elephants also use their trunks for touching, feeding themselves, greeting other elephants, or caressing their young. The trunk can also work as a snorkel if an elephant needs to swim through deep water.

The trunk is absolutely vital for drinking because elephants cannot lower their heads far enough to drink. Though people often assume that elephants use their trunks like a straw, this isn't the case—it's more like a gigantic spoon. The elephant sucks water partway up the trunk, curls it toward its mouth, tilts its head up, and lets the water pour in.

Trunks consist of 40,000 muscles and are incredibly flexible. They can be used for pushing down a tree or picking up something as small as a penny. Trunks get heavy, so it's not uncommon to see an elephant resting its trunk over a tusk.

Other Elephant Characteristics

Tusks are actually the elephant's incisor teeth grown long. Elephants use the tusks to defend themselves and for digging and lifting things. Just as most people are either left- or right-handed, an elephant uses one tusk more than the other.

Elephant teeth are designed to grind plant food, but elephants spend so much time eating that it takes a toll on their teeth, which come in sets of four. When one tooth is worn down, it falls out and another one moves forward. Each elephant can go through six sets of molars in a lifetime. When elephants get old, their teeth are sensitive. If an elephant wears out all its teeth, it can die of malnutrition. Marshes are perfect for older elephants as they can usually find softer plant food there, and many times they stay there until they die. This practice has led to the myth that elephants go to special burial grounds to die. In reality, they are there simply because the area may offer them the best chance of living as long as they can.

The Social Side

Scientists report that elephants seem to have an amazing range of emotions, which include what appears to be the enjoyment of playing as well as sadness and grief. When one elephant has been away from the group and returns, it is greeted with lots of carrying on. During this greeting the elephants will spin around with their heads held high and ears flapping. The ceremonial greeting may also involve an amazing array of sounds ranging from rumbles and screams to roars; they also defecate and urinate with all the excitement.

Elephants also help each other. Elephants have been observed pulling young elephants up muddy banks with their trunks and walking on both sides of weak elephants to help support them. The Walt Disney film *Dumbo* features a host of angry elephants enraged over what happens to Dumbo and his mother, and this isn't far from the truth. If they sense danger, elephants will circle around the babies, flaring their ears to look fierce.

Stories of elephant rage are indeed true, and some scientists feel that elephants suffer post-traumatic stress disorder, which explains their violence when they have been worked too hard or when their food is gone. The tuskless elephant also tends to be more aggressive because it has no way to protect itself.

Elephant Ancestors

Prior to the last ice age, more species of elephants existed than we have today. The most well known of these early species was the mammoth, which inhabited Europe, northern Asia, and North America. Mammoths were more closely related to Asian elephants, and though the term *mammoth* has come to mean "very large," mammoths weren't any larger than today's Asian elephant—the smaller of the elephants we know. Those that lived in the northern areas did, however, have hair and very long, curved tusks. Most mammoths died out at the end of the last ice age; however, a dwarf mammoth that lived on Wrangel Island did not become extinct until about 2000 BC.

While preserved frozen remains of the woolly mammoths have been found in the northern parts of Siberia, they were quite decayed before they became frozen, impairing the ability of scientists to extract much genetic material.

Scientists who look even farther back genetically believe that the elephant family shares distant ancestry with sea cows and hyraxes. One theory suggests that these animals lived underwater and used their trunks as snorkels. Perhaps giving credence to this theory is the fact that modern elephants can swim for up to six hours using their trunks in this manner.

Elephant Families

Both African and Asian elephants live in predominantly female herds, usually consisting of 30–50 animals. The oldest and most experienced female in the herd becomes herd leader, and she decides when and where the herd will eat, rest, and travel. Adult males, called bulls, don't live in herds. Once male elephants become teenagers, they leave their families and live with small groups of other males. After they become adults, they visit other herds for short periods of time to breed. Bulls do not take part in caring for the young.

Within the female herd, family support is everywhere. Elephants bear young only once every few years, and gestation takes twenty to twenty-two months. Once a baby is born, mother elephants receive help from aunts, sisters, and cousins, who are known as *allomothers*. In the process, younger females learn how to care for babies.

Baby elephants are about 3 feet (1 meter) at birth, and about thirty minutes after being born, they struggle to their feet. At first, the baby supports itself by leaning on the mother's leg, but within a few days, it will be strong enough to walk along behind her. Babies are usually quite hairy with long tails and very short trunks that they cannot yet control—baby elephants commonly trip over their trunks until they gain some control over the muscles. Babies find trunks useful, however, and may suck on them like pacifiers. Babies nurse exclusively for about two years and continue to nurse for up to six years. They learn primarily by observation, not from instinct.

Adolescent males determine their own ranking order through jousting contests using head and tusks. Strength and temperament are as important as size and age. Unless a female is in estrus, males tend to be very tolerant of each other. When a female is ready to be impregnated, the males become competitive with one another.

When elephant families are destroyed by poachers, they suffer a disorder such as you might expect when human families are destroyed, scientists are discovering. In an article in *Natural History* magazine (Delia Owens and Mark Owens, "Comeback Kids," July/August 2005, 22–25), wildlife specialists Delia and Mark Owens relate their observations of what has happened in Zambia where poachers shot 93 percent of the elephants in the population for their ivory, skin, or meat. The population declined from 17,000 in the mid-1970s to approximately 1,500 in the mid-1990s. After identifying a five-year-old female orphaned elephant, the Owenses carefully tracked her development. She gave birth at a much younger age than most elephants, and because she had not had the protective herd experience herself, she all but neglected the baby once it was born to her. (By age sixteen, the age when she would normally have given birth to her first calf, she had had three calves and one grandcalf.) The Owenses managed to follow her (they called her Gift) long enough to observe more offspring being born. As the group grew, Gift became more of a matriarch and did better at tending to others. Still, the social disruption is devastating to the family groups and to the species as a whole.

Elephant Communications

Elephants employ a variety of squeaks, gurgles, trumpeting sounds, snorts, barks, and rumbles to communicate, using more than 200 different sounds, including some infrasonic sounds that are too low for a human to hear. The low, resounding calls elephants make can be heard by another elephant up to 5 miles (8 kilometers) away, and these long-distance sounds may be how female elephants let males know they are ready to mate. (Females can become pregnant only a few days every year, so timing is everything.)

Interestingly, stomach growls, which are loud enough for other elephants to hear, are a welcome sound to elephants and seem to mean "everything is okay." Scientists have also been able to document the elephant's ability to mimic other sounds, which probably explains the range of the noises they make. Two elephants who lived in captivity proved adept at learning new sounds. One case concerned an African elephant orphan who was being raised at an orphanage in Kenya. The night stockade where the elephants stayed was 3 kilometers from the Nairobi-Mombasa highway, and each evening the adolescent female would make a calling noise that mimicked truck and traffic noise.

Another case was in a Swiss zoo where a twenty-three-year-old male African elephant lived with two Asian elephants.

Over time he came to mimic the chirpy calls that are typical of Asian elephants.

The Domestication of the Elephant

India may have been the first country to domesticate the elephant. Elephants have been used for heavy labor (uprooting trees and moving logs) and safaris and as ceremonial mounts as well as in war. Asian elephants are easier to domesticate than African ones. Male elephants, however, are not that easy to train because they periodically go into a state known as *musth*, when they can be quite difficult to control. Females are easier to train, but warriors soon found that they weren't helpful in battles as they will run if they encounter a male elephant.

Hannibal, the Carthaginian general, took elephants across the Alps when he was fighting the Romans. Because domesticating a large number of elephants was never feasible, armies often held their elephants in reserve to be brought forward only for the most important battles. These great beasts offered great confidence for those warriors who possessed them, and they inspired great fear in the opposing men and often spooked the opposing army's horses or camels.

In some cultures, elephants were equipped with frightening headpieces and breastplates as defensive armor. Spears could be attached to the tusks to increase the amount of damage a charging elephant could do. Sometimes an elephant went forward with only its trainer (mahout, usually from India) on board; other times, it carried several soldiers. Sometimes, even a tower or castle (called a *howdah*) with a crew of three soldiers with arrows or long pikes was fastened to an elephant's back for fighting.

Until gunpowder came into use in the late fifteenth century and made it easier to bring down an elephant, these massive animals were the early fighters' equivalent of heavy tanks. They could be used to bash through obstructions and certainly succeeded in terrorizing those in their path. Elephants have also been employed as executioners and were taught to crush the condemned with their feet.

The first domesticated elephant to come to the United States was a two-year-old elephant brought to New York by a Captain Jacob Crowninshield in 1796. He purchased the animal in India, and once in the United States, it was toured around the country as the "greatest natural curiosity ever presented to the public." People paid between twenty-five and fifty cents to see it.

As late as World War I, the elephant was still used for pulling heavy equipment.

Enemies

Elephants have few natural enemies. Crocodiles and lions can prey on young or weak elephants, but otherwise, it is poaching (illegal hunting) that has brought African elephants to the brink of extinction. In 1970 there were 2,000,000 African elephants; in 2000 there were between 400,000 and 600,000. Asian elephants breed well enough in captivity, so their numbers have remained more stable.

Unfortunately, elephants have been relentlessly hunted for their tusks. Hunters must kill the elephants because one-third of the tusk is inside the head. To maximize value in a land of poverty, the poachers want all they can get.

What's Being Done to Protect the Elephants?

The tusk is an important tool for the elephant as it uses its tusks to spar, tear apart vegetation, and root around in the ground for necessary minerals. (Like some other herbivores, elephants seek out natural mineral licks and consume a certain amount of soil; scientists speculate that they are doing this primarily when it is necessary to increase their intake of sodium.) Unfortunately elephant anatomy has been affected by all the poaching.

By killing only tusked elephants there has been an increased opportunity for elephants with small tusks or no tusks to mate. (Only the male Asian elephants have tusks; females have *tushes,* second incisors that stick out beyond the upper lip.) Elephants with no tusks used to be a rare genetic abnormality. However, today in some populations of African elephants almost 30 percent of the animals are tuskless, compared with 1 percent in 1930. Now it's becoming a hereditary trait (and a clear demonstration of evolutionary change through natural selection). While this type of change normally takes thousands of years of evolution, with elephants it seems to be happening faster. And without tusks elephants will likely also have to change their behavior.

The elephant is now a protected animal, and more than twenty years ago the Convention on International Trade in Endangered Species (CITES) banned the sale of ivory starting in 1989. Elephant populations started to come back a bit, but in 1997, three countries were allowed to resume trade, selling ivory seized from illegal hunters. While the money gained from these sales goes into a fund to help elephants and elephant land preserves, others feel that it has led to a renewed interest in poaching.

The Internet May Undo the Progress Made

Despite the ban on ivory trade, it is very difficult to fight the illegal wildlife and wildlife parts trade. As of August 2005, the International Fund for Animal Welfare in Yarmouth Port, Massachusetts, reports that it has found more than 6,000 illegal or potentially illegal wildlife items for sale online, including an elephant-bone sculpture for $18,000.

Endangered species are covered by a complex web of local, state, federal, and international laws and treaties, which include a number of exceptions, making it all the more complicated for anyone to police activity. What's more, the Fish and Wildlife Service, the governmental body responsible for enforcing most federal animal protection laws, has only 219 agents. To halt Internet activity, agents often need to seize computers and search carefully through pages and pages of e-mail to document illegal trade.

Animal welfare advocates say that sites such as eBay and Overstock.com are helping by carefully monitoring what is sold by their vendors, and perhaps this will help slow illegal activity. In the meantime, we can all do our part by keeping in mind that buying ivory jewelry or animal skins or items made from animal parts is keeping the illegal animal trade alive.

In an effort to further understand the elephants to save them, scientists with the Elephant Research Project in Kenya's Amboseli Park are now tagging the elephants with global positioning satellite receivers to allow the use of satellite imagery to track the elephants.

And in India scientists have been working to fight habitat fragmentation. If elephants can't move freely, it reduces their ability to find food and promotes inbreeding, which is not good for any species. Scientists are trying to set aside 20-mile "paths" so that elephants can move from habitat to habitat. Dr. Raman Sukumar, a leading ecologist and the director of the Asian Elephant Research Centre in Bangalore, India, who recently received a prestigious award from the Wildlife Trust for his work in studying the Asian elephant, noted at a luncheon given for him by the Wildlife Trust in May of 2005, "In saving the elephant, we save ourselves."

If we're able to maintain the biodiversity necessary to keep elephants alive, we will, in the process, save our world. The environment necessary for healthy elephants is one that is healthy for other plants and animals as well as humankind.

5

Mosquitoes

Annoying Pest and Deadly Foe

In the autumn of 1999 the facts about a medical mystery unfolded like a serialized story on the front page of New York metropolitan newspapers as new developments concerning a new mosquito-borne illness were gradually revealed. Large numbers of birds in the area died, and smaller numbers of people (mostly the elderly) were sickened, and a few of them died as well. Medical specialists from the Centers for Disease Control and Prevention (CDC), the U.S. Army Medical Research Institute for Infectious Diseases, and a pathologist at the Bronx Zoo worked to unravel the mystery of what was going on.

When specialists at the CDC were first contacted about the human deaths in the New York City area, they originally tested the samples against diseases that were known to already exist in the United States. A close match was found with St. Louis encephalitis, a mosquito-borne illness that crops up in this country from time to time. However, as Dr. Tracey S. McNamara, head of pathology for the Wildlife Conservation Society (Bronx Zoo), thought about it, this began to puzzle her. Shortly before these deaths were reported, she had noticed an unusual number of birds dying in and around the zoo. Reports from neighboring communities of the death of crows in the area made her begin to think that the coincidence between the bird deaths and the human deaths was too great to ignore.

After running tests on bird samples at the National Veterinary Services Laboratory in Ames, Iowa, and prevailing upon a contact at the U.S. Army lab to run some additional tests for her, Dr. McNamara was able to approach the CDC again with the specifics. It wasn't the St. Louis encephalitis as originally thought; it was a virus never before seen in the western hemisphere, the West Nile virus.

As of 2005, the Centers for Disease Control and Prevention has counted more than 16,600 human cases and 654 deaths nationwide. While it was originally thought that the disease would be only a mild illness in most sufferers, the CDC now reports that in 2004, approximately one-third of the West Nile cases reported had neurologic complications (encephalitis or meningitis). Some patients come down with a polio-like paralysis or severe muscle weakness. A 2004 study by Chicago's health department found that half the people who had West Nile were sick enough to stay out of school or work for ten days and suffered fatigue that lasted a month. The median time to get back to normal was sixty days.

All this resulted from a mosquito-borne disease that first appeared in Uganda in 1937 and we had never seen in this country before 1999.

More than a seasonal annoyance, mosquitoes transmit dozens of devastating diseases, and these insects are, indeed, more than a pesky outdoor nuisance. In a day when the medical community has completely eradicated smallpox, it seems amazing that a tiny insect can be so effective at making people so ill—and West Nile is only part of the mosquito story. According to the World Health Organization, some 500 million people are infected with mosquito-borne illnesses each year (malaria, dengue, yellow fever, and various forms of encephalitis, including West Nile), and more than 2.5 million of them die, many of them young children.

The Mosquito

Insects have been around for at least 400 million years, and over time, they have diversified into an estimated 5 million living species, dwarfing the diversity of all other animals combined. They survived well through the mass extinctions that affected the dinosaurs and marine reptiles 65 million years ago. Scientists report that today there are approximately 2,700 species of mosquito, and they are arranged in approximately 34 genera.

While this cartoon pokes fun at the size and abundance of mosquitoes in Nome, Alaska, the insect's role as purveyor of disease is anything but humorous. Source: Library of Congress.

We all know basically what a mosquito looks like, but what you probably don't know is that mosquitoes weigh about 2.5 milligrams and can fly at a rate of 1.5 miles per hour. (Millions of years ago, mosquitoes were three times the size they are today!) Mosquitoes have two vibrating knobs on short stalks that stick out of their bodies like pins, one right behind each wing. These are called *halters*, and they help keep the mosquito balanced in flight.

To find its victims, the mosquito's antennae have thousands of hairs that sense moisture, lactic acid, carbon dioxide, body heat, and movement. The biting mosquito is always a female; only females feed on blood. Once they make their bite, the blood triggers ovarian activity, and after a few days your blood will have permitted her to lay several hundred eggs. (Mosquitoes can lay eggs almost anywhere they find at least one inch of standing water.) A few weeks later, the eggs hatch. Don't bother to wonder about mosquito parents dealing with toddlerhood or adolescence—two or three days later, these "babies" become full adult mosquitoes.

While we are always taught to respect nature because insects "do important work," this is not true for the mosquito. They don't pollinate plants, they don't aerate soil, and they don't spread seeds, and they aren't even beautiful. Suffice it to say that mosquitoes basically have no redeeming virtues. They exist as a "miracle of nature" and to make more little mosquitoes. Smack! Smack!

Why We Should Be Concerned

Arbovirus is the term used to describe arthropod-borne viruses—viruses that are transmitted to birds and vertebrate hosts (animals with backbones) by blood-feeding insects such as mosquitoes. (Arthropods are invertebrates, such as insects, spiders, and crustaceans, that have segmented bodies and jointed appendages.) Mosquito-borne diseases are spreading, partly because of our encroachment on their habitats, and partly because of the warming of the planet—disease-bearing mosquitoes can move into temperate regions that were previously too cold for them.

While no one is claiming to have located the first mosquito to arrive in the United States carrying West Nile, what they do know now is something about the ways in which West Nile—and other arboviruses—can travel the globe. One rather surprising finding from a 1987 study is that mosquitoes can survive in the wheel bays of jets traveling thousands of miles between tropical and temperate zones. While "leaving on a jet plane" is one manner in which disease-carrying mosquitoes may make some transcontinental leaps, there is also the tried-and-true bite-and-flee method (the insect bites a host and flees, but the host transports the disease when he or she travels to another part of the world), and the arboviruses benefit from our mobile world society.

Arboviruses require a host (usually a bird or small mammal) in which they maintain themselves in nature, and a vector, such as a mosquito, to get around to infect other organisms. Female mosquitoes may ingest a virus from an infected host and later pass the infection in their saliva when they bite another animal. Keeping in mind that in every country there are many types of mosquitoes, here's a likely and understandable scenario as to how an infection might spread transcontinentally:

A businessman is in Venezuela for a meeting, and he decides to take an evening stroll. While out on the stroll, he is bitten by an *Aedes aegypti* mosquito. The next day his business is concluded, and he flies home to Arizona. That weekend he isn't feeling terrific, but the family is coming over for a barbecue, so he is outside preparing to grill hamburgers when he is bitten by an Arizona mosquito, who picks up the dengue fever virus and begins a cycle that will permit it to spread through Arizona.

The Importance of Being "On Watch"

The arrival of a disease that had never before been seen in the western hemisphere had the result of making people in the United States aware that they needed to be on watch for other mosquito-born illnesses. Here are some that are around:

Dengue Fever

Dengue fever is generally a tropical disease that involves fever, bad headache, joint aches that are quite severe, a rash, and in the more severe form, a hemorrhagic component.

The dangers of DDT are so severe that no one can wish for it to be brought back; however, during the time it was used there were some benefits. Dengue fever was nearly eradicated because of it, and in the 1950s, the World Health Organization was seeing an annual count of fewer than 1,000 cases of the sometimes fatal hemorrhagic variety. Then DDT was taken off the market (see chapter 15), and the mosquito returned with a vengeance. As a result, dengue fever has gone from a nonissue to an exploding one, with the cases up 400 percent from the 1970s. In 2000 there were more than 500,000 cases worldwide. In the United States, the main carrier of dengue is the *Aedes aegypti* mosquito, and the disease has thus far been diagnosed in eleven states from Arizona to North Carolina.

Eastern Equine Encephalitis

Eastern equine encephalitis appears cyclically in the United States. The virus is passed from mosquitoes to birds and then from birds to mosquitoes to humans. EEE is the most serious type of encephalitis, a brain inflammation, and it can't be treated. It is fatal in 50–75 percent of cases.

Rift Valley Fever

Rift Valley fever has not yet reached the United States. Most sufferers of RVF have flulike symptoms and may develop more serious problems such as eye, liver, or kidney disease. In the 1930s, RVF killed tens of thousands of sheep and also spread to humans. In Saudi Arabia in 2000, Rift Valley fever killed about 100 people and sickened another 800.

What Our Global Consciences Should Remember

While the United States worries about West Nile or the avian flu, what we sometimes fail to acknowledge is the number of recurring infectious diseases that claim a high toll in less-developed countries every single year. For example, tuberculosis kills an estimated 2 million people each year, and approximately 40 percent of the world's population is at risk for malaria, a mosquito-borne illness, which causes 300 million acute illnesses and at least 1,000,000 deaths annually. (These figures dwarf seasonal influenza, which may cause up to 5 million cases of severe illnesses and between 250,000 and 500,000 deaths every year.) It imposes $12 billion a year in economic costs on sub-Saharan Africa, the hardest-hit region.

Malaria is known as a "relapsing fever" because of the way the virus develops. (The word *malaria* comes from the Italian words for "bad air," because people used to think that this illness was caused by the air from marshy or swampy lands.) It is caused by a parasitic protozoan that causes symptoms of fever, shivering, joint pain, vomiting, anemia (caused by the destruction of red blood cells), and convulsions.

The infection spreads when the microbe enters its host via the parasite-laden saliva of a female *Anopheles* mosquito. Minutes after she inserts her proboscis, organisms are released that then travel through the host's blood on their way to the liver, where they multiply. Seven to ten days later, 10,000–30,000 descendants return to the bloodstream, each ready to raid a red blood cell and siphon hemoglobin to fuel the birth of another 10–20 "babies" per cell. Forty-eight hours later, the next stage of the illness occurs—malaria's infamous fevers and chills. The cycle of transmission is sustained when a new mosquito bites an infected person and the disease spreads.

Some people have genetic mutations that hinder the ability of the parasite to grow within the red blood cells, and for that reason they don't get sick—or *as* sick. Very young children are more likely to die from malaria because they have not yet build up any immunity.

Effective methods of prevention include spraying insecticides on interior walls and covering sleeping children with insecticide-treated mosquito nets. Drugs can also be used to treat the disease. For many years, some affordable antimalarial drugs such as chloroquine proved to be quite effective; however, today, due to overuse, these drugs have lost their punch. Now a more expensive combination drug is needed; the core ingredient is artemisinin, a plant extract.

International health agencies launched a Roll Back Malaria campaign in 1998, with the goal of cutting malaria deaths in half by 2010. Progress, however, has been slow. The Gates Foundation stepped forward and has provided major funding to the Global Fund to Fight AIDS, Tuberculosis and Malaria. Zambia has been the prime focus for the Gates outreach. In Zambia, 30,000 people per year die, and the goal of the outreach is to cut malaria deaths by 75 percent in three years. The foundation felt that the Zambian government was doing the right things but lacked the funding to follow through.

Technology versus the Mighty Mosquito

In late November 1905, a biotech company announced that it had created a "smart" mosquito net, a wireless technology that could kill mosquitoes.

If You Like Math, You Can Help Fight Malaria

In studies conducted in malaria-prone countries, scientists have determined that 20 percent of the children get 80 percent of the exposure to malarial mosquitoes. Researchers believe that understanding a mathematical model based on these figures is key to controlling the deadly disease.

Some people are bitten more often because they live where mosquitoes are more common or because the mosquitoes find them more attractive, and those who are bitten most often play a role in malarial transmission similar to that played by the most sexually active in transmission of STDs. In other words, the heaviest burden of the disease is carried by a small minority of people. (People who are immune can still be carriers.) The challenge is identifying the most frequently bitten.

It is hoped that by focusing on the small number of people who are bitten most often, scientists will be able to reduce the number of asymptomatic carriers, lower the rate of parasites among mosquitoes, and, finally, decrease overall transmission among the population.

The description of the machine is as follows: "It emits an odorless formula that combines carbon dioxide, moisture, heat, and a short-range attractant called *octenol*, and each magnet, powered 24 hours a day by propane or electricity, pulls bugs in from as far away as 100 yards" (Stefanie Olsen, "Wi-Fi Mosquito Killer Coming to a Porch Near You," CNET News.com, November 21, 2005). It works like a vacuum cleaner that dehydrates and kills them. The biotech company servers are programmed to take in information from out there in Mosquito Land, so that the equipment can shut down during rainstorms, or if the wind is blowing hard enough to simply blow the mosquitoes into another area.

Despite the amount of time and effort that must have gone into creating this magical electronic mosquito magnet machine, a professor of entomology says that what really works is floating a harmless oil on the ponds so that the larvae suffocate. The professor notes, "You win the war by keeping down the numbers of soldiers" (Olsen).

PART TWO

Scientific Discoveries That Have Changed Our Lives

6

How They Learned Why We Get Sick

The Origin of Germ Theory

In today's world, even very young children know that we get sick because of germs—children who go to school with a bad cold know they may pass that cold on to someone else. Knowledge of germs also has improved treatment: Added to our age-old remedy of "stay home and take care of yourself," we have two medical avenues available to us to combat disease—antibiotics to fight off bacterial illnesses, and vaccines to prevent some illnesses (chicken pox, measles, flu, etc.) from striking at all.

Because we feel we know so much about sickness, it is difficult to comprehend that only 150 years ago, people didn't have any idea about what made them ill. It took until the 1860s and Louis Pasteur's discovery of germs before the human race began to understand what makes people sick.

Pasteur's discovery created a great turnaround in the field of medicine. Once germs were identified, vaccines were created and treatments sought. His discovery also led the way for a big improvement in public health in Europe.

"If it is a terrifying thought that life is at the mercy of the multiplication of these minute bodies, it is a consoling hope that Science will not always remain powerless before such enemies."

—LOUIS PASTEUR, paper read before the French Academy of Sciences, April 29, 1878

Louis Pasteur.
Source: Library of Congress.

Early Ideas about Illness

Early on, people thought they knew why we become ill. The Egyptians believed that it had to do with the body's "channels," and Greeks and Romans felt that every person consisted of four humors; when someone became sick, it was because the humors were out of balance. In 400 BC the Greek physician Hippocrates (of Hippocratic oath fame) put forth the theory that disease came from smelly gases. Well into the 1800s physicians continued to feel that miasmas, gases that arise from sewers, swamps, garbage pits, or open graves and thought to be poisonous, caused illnesses.

When it came to treatments derived before germs were recognized, guesswork was heavily involved. Doctors frequently used leeches to bleed patients, and laxatives, opium, peppermint, and brandy were often considered cures. While some herbal medications have come back into favor today, many of the medicines used early on—mercury among them—are now known to be poisonous or cause serious, if not fatal, damage.

First Sightings of Bacteria

As always, some important discoveries preceded the scientist who achieved the breakthrough. Though microscopes were actually invented by scientists who were working late in the 1500s, a Dutch cloth merchant actually discovered bacteria.

Antony van Leeuwenhoek (1632–1723) was an unlikely scientist. A tradesman born into a family of tradesmen, he lived in Delft, Holland, and though he worked as a fabric merchant, he was intensely curious and had an all-consuming hobby. He was fascinated by what was beyond the vision of the naked eye. He ground his own lenses and made over five hundred microscopes. Though these were simple devices compared to what we have today, Leeuwenhoek was able to create a more powerful microscope than the compound microscopes created by Robert Hooke in England and Jan Swammerdam in the Netherlands, which magnified things only 20 to 30 times. Leeuwenhoek's grinding skill, his acute eyesight, and his intuitive understanding of how to direct light onto the object permitted him to see items that were estimated to be 200 times their natural size.

Leeuwenhoek studied animal and plant tissues as well as mineral crystals and fossils; he was the first to see microscopic animals such as nematodes (roundworms) and rotifers (multicelled animals that have a disk at one end with circles of strong cilia that often look like spinning wheels) as well as blood cells and living sperm. Leeuwenhoek created careful descriptions of exactly what he saw, and he hired an illustrator to draw what he described. He began submitting his information to the Royal Society of London, and because of his meticulous efforts—as well as his unique discoveries—the Society overcame their preference for work from credentialed scientists and recognized the merit of this cloth merchant's work. They had his descriptions translated from Dutch into English or Latin, and his findings were regularly published in the Society's publication.

The Little Animalcules in Tooth Plaque

In 1683, Leeuwenhoek wrote to the Royal Society about his observations of the content of tooth plaque. (In those days, tooth brushing would not have occurred regularly—if at all—as no one would have recognized the advisability of clean teeth.) He wrote,

> **I then most always saw, with great wonder, that in the said matter there were many very little living animalcules, very prettily a-moving. The biggest sort . . . had a very strong and swift motion, and shot through the water (or spittle) like a pike does through the water. The second sort . . . oft-times spun round like a top . . . and these were far more in number. (Brian J. Ford and Al Shinn, *Antony Van Leeuwenhoek,* University of California at Berkeley Museum of Paleontology, www.ucmp.berkeley.edu/history/leeuwenhoek.html. Retrieved November 16, 2005)**

In the mouth of one of the old men whose plaque he studied, Leeuwenhoek found "an unbelievably great company of living animalcules, a-swimming more nimbly than any I had ever seen up to this time. The biggest sort . . . bent their body into curves in going forwards. . . . Moreover, the other animalcules were in such enormous numbers, that all the water . . . seemed to be alive" (Ford and Shinn).

These were among the first observations on living bacteria ever recorded.

From Frogs to Germs: The Belief in Spontaneous Generation

Pre-Pasteur, "spontaneous generation" was believed to explain the surprising appearance of living things where only nonliving things had previously existed. For example, when the Nile River in Egypt flooded each spring, nutrient-rich mud covered the river banks, and soon this fertile land along the water's edge was filled with frogs. The Egyptians concluded that muddy soil gave rise to frogs. Later on, medieval European farmers used to store grain in barns with thatched roofs that often leaked, and the grain became moldy. Mice, of course, hovered around the grain-filled areas, so the belief arose that mice came from moldy grain.

Because microscopes weren't invented until the 1600s—

and even then they weren't very powerful—scientists couldn't even imagine something that was invisible to the naked eye. To them, spontaneous generation of living things from nonliving things was the only explanation that made sense.

By the nineteenth century, spontaneous generation was hotly debated. People were beginning to decide that larger organisms didn't generate spontaneously but that smaller ones still did. Because the topic was very much under discussion, the Paris Academy of Sciences offered a prize for any experiments that would help resolve the conflict. In 1864 the prize was awarded to Louis Pasteur for experiments that definitively proved that microorganisms are present in air but that air cannot give rise to organisms spontaneously.

Pasteur Accomplished More Than Pasteurization

Louis Pasteur (1822–1895) was born in Jura, France. While training to be a teacher at a school established in Paris by Emperor Napoleon III, he showed an early aptitude for chemistry. Though he eventually became a professor at the University of Strasbourg, his work with molds and his studies of crystals were known to Napoleon, who asked him to step in to try to help the ailing French wine industry. After studying the fermentation process, Pasteur eventually found that if the wine was heated to 55 degrees Celsius for several minutes, then the microorganisms in it were killed and the wine didn't spoil—a process we now know as *pasteurization.* From this work, Pasteur was able to show that decay in organic matter was caused by germs—what we now know as *microbes*—that floated in the air. Ultimately, Pasteur went on to demonstrate by experiment that microscopic organisms could cause illnesses.

German microbiologist Robert Koch built on this information and set out three laws (1883) that explained the cause of disease. Koch's postulates have been used ever since to determine whether an organism causes a disease:

1. The suspected germ must be consistently associated with the disease.
2. It must be isolated from the sick person and cultured in the laboratory.
3. Experimental inoculation with the organism must cause the symptoms of the disease to appear.

In 1905 a fourth rule was added:

4. Organisms must be isolated again from the experimental infection.

Using Pasteur's theory and Koch's postulates, scientists have figured out cures for disease after disease. Pasteur's germ theory became the foundation of the science of microbiology and a cornerstone of modern medicine.

As Pasteur continued to experiment, he learned that different microbes caused different diseases and found that with some illnesses, a weakened form of the microbe could be used to immunize against more virulent forms. His work with rabies was particularly important. After finding that rabies was transmitted by a virus, Pasteur developed both an effective treatment—in 1885 he saved the lives of two boys who had been bitten by a rabid dog—and learned that he could also effectively vaccinate dogs themselves against rabies. He went on to develop vaccines against chicken cholera, anthrax, and swine fever.

In 1886 the Pasteur Institute was founded in France as a clinic for rabies treatment, a research center for disease, and a teaching institute. When a Pasteur Institute was founded in Saigon in 1891, it became the first in a world network and showed the esteem in which his work had eventually come to be held.

The Curative Power of Cleanliness

Medical knowledge in the mid-nineteenth century was desperately inadequate. To put the situation in perspective, it is

John Snow's Fight for Clean Water in England

During the nineteenth century, thousands of people were dying of cholera. England's Dr. John Snow (1813–1858), who later became known as the Father of Epidemiology, felt that this acute intestinal illness was caused by people drinking dirty water. Despite his effort, he was unable to come up with scientific proof, which could have forced the government to upgrade public sanitation and water systems.

However, once Pasteur made his discoveries, Snow's work was viewed in a new light. It became clear that it was very likely that the cholera germ was spread by water that had been contaminated by sewage and rubbish. As a result, Britain passed the Public Health Act in 1871, which forced councils to remove sewage and rubbish from the street and to provide communities with clean drinking water. Overall health improved.

Snow died in 1858 before his mission was accomplished.

---

interesting to note that in the United States during the Civil War, fully two-thirds of the 618,000 casualties from the war were from disease (typhus, pneumonia, infectious diarrhea, typhoid, and tetanus), not from battle wounds. Part of the spread of disease was due to unsanitary conditions. In hospitals, surgery was performed without gloves, and instruments were wiped clean on the physicians' aprons.

With this in mind, it's easier to understand the import of what a Hungarian doctor, Ignaz Semmelweis, discovered when practicing obstetrics at the Allgemeine Krankenhaus (Vienna), starting in the 1840s. At the hospital, it was common practice for doctors to do autopsies in the morning and perform pelvic examinations on expectant women or deliver babies in the afternoon. No one knew about sterilization of instruments or the importance of washing hands or wearing gloves, and puerperal fever (childbirth fever) was rampant.

Dr. Semmelweis thought uncleanliness might be to blame and ordered the doctors to wash the pus, blood, and tissue from their hands after the autopsies and before seeing patients. Deaths from infections on Semmelweis's wards plunged (from 12 percent to 1 percent), but because Semmelweis, an abrasive personality who felt "his say" should be enough, lacked a way to prove his ideas, they were met with resistance.

A few years later, Joseph Lister, a Scottish physician, made the link between Pasteur's discovery and what Semmelweis had accomplished through his insistence on hand washing. Lister implemented sanitary practices within hospitals, including careful wound cleaning as well as a much cleaner operating environment.

Be Glad You Live in a Modern Nation

We read about Semmelweis and what he learned about the importance of hand washing, and we assume that now "everyone knows that." Not so. Every year more than 3.5 million children under the age of five in low-income communities in developing countries die from diarrhea and acute lower respiratory infection.

Researchers at the U.S. Centers for Disease Control and Prevention undertook a study in Karachi, Pakistan, recently to find out what effect hand washing would have on the occurrence of illness. Establishing a randomized control trial, they entered twenty-five neighborhoods and provided either plain soap or antibacterial soap and promoted hand washing; eleven other neighborhoods served as controls. Children under five in households that received plain soap had a 50 percent lower incidence of pneumonia than the control groups. Children under fifteen in households encouraged to use plain soap had a 53 percent lower incidence of diarrhea and a 34 percent lower incidence of impetigo. There was no significant difference in disease levels between those given plain soap and those given antibacterial soap. In this study, respiratory infections did not decrease immediately after the introduction of hand washing but fell substantially after some months.

So though it seems that illnesses caused by uncleanliness should have come to an end more than a century ago, it is obvious that in some places in the world there are simple lessons yet to be learned.

Other Helpful Developments

Though there is still much to learn, medical science has clipped along at a great rate, and some of the particularly significant developments include the following:

- In 1892 Russian scientist Dmitri Ivanovski proved that there are disease-causing organisms smaller than bacteria, which became known as viruses. It took the invention of the electron microscope in the 1930s for scientists to see them.

- In 1928 Scottish bacteriologist Alexander Fleming noticed that a mold called *Penicillium* was effective at killing bacteria. This led to the development of the first antibiotic. British biochemist Ernst Chain and Australian pathologist Howard Florey began to mass-produce penicillin during World War II. This discovery allowed treatment of previously fatal illnesses, such as tuberculosis, all because of Pasteur's original discoveries.
- In 1946 the United States formed the Communicable Disease Center (now Centers for Disease Control and Prevention), the nation's disease watchdog that has helped lead worldwide battles against smallpox, polio, and other illnesses.
- In 1980 a global vaccination program led to eradication of the virus that causes smallpox. This is the only time a germ has successfully been made extinct. (Some countries have kept samples of the virus, which offers a frightening prospect if it is ever used in biological warfare.)

What We Are Learning Now

Scientists today are learning that disease is much more complex than has been suspected. They now realize that a lot of chronic illnesses that were thought to be genetic or environmental may turn out to be caused by bacteria or groups of bacteria or viruses. An April 2005 report from the American Academy of Microbiology (as quoted in Nicholas Bakalar, "More Diseases Pinned on Old Culprits: Germs," *New York Times*, May 17, 2005) indicates that, for example, diabetes, never before suspected of being a microbe-caused illness, may one day turn out to be a complex immune reaction to a previous infection. Recently an Australian ophthalmologist made a connection between children with cataracts and mothers who had German measles during pregnancy (Bakalar). Just as the varicella virus (which causes chickenpox) is known to lurk in the body and reappear to cause herpes zoster or shingles at times when the body is weakened, we may one day discover that a lot of chronic illnesses thought to be genetic or environmental may have at their roots a bacteria or groups of bacteria or viruses.

Stay tuned.

7

Darwin

Then and Now

For a fellow who died more than one hundred years ago, Charles Darwin (1809–1882) has had an overly generous share of headlines in the early twenty-first century, and no one would be more upset by the fuss than he. Darwin was a very unlikely candidate to become the originator of an idea that was to turn the scientific world upside down.

As revolutionary as the ideas of Galileo, Newton, and Einstein, Darwin's explanation of evolution by "natural selection" transformed our understanding of the living world, and his theory of natural selection underlies all of modern biology.

Biology's Debt to Serendipity

Charles Darwin was born into a wealthy family. His advanced education began with the study of medicine, but when Darwin found himself uncomfortable with the thought of conducting surgery, he switched to major in divinity. His interest in the natural world was an actively pursued hobby, and while in school he participated in student societies that united others with similar interests in nature.

We have serendipity to thank for the fact that Darwin did not return home to become a vicar after he graduated. Several incidents occurred that were to provide him with the opportu-

"There is a grandeur in this view of life . . . from so simple a beginning, endless forms most beautiful and most wonderful have been, and are being, evolved" (final paragraph of *The Origin of Species*, 1859).

It Wasn't the Concept of Evolution That Was New . . .

Evolution was widely discussed among scientists prior to Darwin. Suggested by zoologist Jean-Baptiste Lamarck (1744–1829), evolution was also supported by Erasmus Darwin (1731–1802), Darwin's grandfather and a well-respected doctor and botanist. However, the idea did not hold weight for most people in the nineteenth century because there was no explanation for how species could evolve into anything but what they already were. In addition to a lack of scientific evidence, the theology of the day was that a beneficent God oversaw the creation of everything—things didn't just evolve. Early in his life, Darwin agreed with the argument that a deity had to have arranged life, that it was so complex that it could not have "just happened."

Darwin is a household name because he changed his mind and then found and proved a viable mechanism for evolution, namely natural selection.

nity and the background to develop the theories for which he is known today.

After graduating in 1831 at the age of twenty-two, Darwin was invited to accompany twenty-three-year-old Captain Robert FitzRoy as a "gentleman companion" on a two-year journey on FitzRoy's assigned ship, the *Beagle.* FitzRoy's mission was to chart the ocean waters of South America and return via New Zealand. Ironically, FitzRoy's hobby was collecting evidence from nature that would support the biblical interpretation of creation.

Darwin's qualification for this position consisted of being available and being born into the proper social class. In those days, exploratory expeditions were so long that captains were known to become very lonely and depressed. They were of a different social class from the crew, so it was common practice to take along friend—or someone of good social standing—as a companion. For this voyage, FitzRoy's first choice for the role had backed out, so Darwin was his replacement.

In another bit of serendipity, Charles Lyell, considered the father of geology, whom you meet later in this book, had asked FitzRoy to record geological features he observed on his travels. Lyell provided FitzRoy with a copy of his first volume of *Principles of Geology,* which FitzRoy then gave to Darwin. Lyell's book expounds upon his belief that the earth was created grad-

Charles Darwin.
Source: Library of Congress.

ually over an extremely long span of time—much longer than the several thousand years most theologians of the day believed. This concept of the earth's timeline had to have been instrumental to Darwin's conceptualizing natural selection.

The Trip

Once the voyage was underway, Darwin made excellent use of what was to become a five-year journey. The role of ship's naturalist was usually filled by the ship's surgeon, and this was true for the voyage of the *Beagle* as well. However, as fate would have it, the *Beagle*'s surgeon left the ship at a port along the way, so Darwin was left in charge of collecting specimens, which he then arranged to have shipped back to England.

Though the trip held no single moment of illumination for Darwin, there were many significant discoveries that helped shape Darwin's theories, as we can see when we view them collectively and in hindsight. Among them are these:

- Darwin witnessed the eruption of a volcano in the Andes and the immediate aftereffects of an earthquake in Chile. After the Chilean earthquake, Darwin noticed that the ground had risen some 9 feet. He could see that the mussel beds were lifted high above the highest water, in a clear demonstration of the truth behind what Lyell had written.
- Though Darwin is famous for noting the differentiation among what have become known as Darwin's finches, this discovery did not come to Darwin without help. Natives of the Galapagos Islands pointed out to him that both tortoises and finches differed from island to island—you could tell what island you were on based on the tortoise or the finch you found there. This information did not particularly impress Darwin at the time, and he poorly catalogued the finches he sent home. Fortunately, he had an assistant who made careful notations, and this permitted Darwin to rediscover at home what he had not focused on while away.
- Darwin's exposure to the primitive natives of the Tierra del Fuego made Darwin begin to wonder if civilization itself had evolved over time.
- When he saw the armadillos in Argentina, he was struck by their resemblance to fossils he was finding. Darwin wondered why so many species were extinct but were replaced by something strikingly similar. Darwin began to suspect that a grand plan that involved the separate creation of each slightly different species did not make a lot of sense.

As he shipped specimens home, he also sent along reports of all he was seeing. His early reports were assembled by Reverend Professor John Stevens Henslow (1796–1861), who arranged to give certain naturalists access to Darwin's specimens and his reports. As a result, when Darwin returned, he was a celebrity among scientists, respected for the work he was doing.

Once he returned to England in 1836, he never left home again, but there was plenty to keep him busy. Shortly after his return, Captain FitzRoy began assembling a book on the voyage and asked Darwin to contribute the section on the specimens collected from nature. Originally published as part of FitzRoy's account and known as the *Journal of Researches into the Natural History and Geology of the Countries Visited during the Voyage Round the World of H.M.S. Beagle*, it was eventually published on its own and became known as *The Voyage of the Beagle*. Darwin continued to revise his writings, and by 1845, his revisions began to explore the idea of natural selection.

Much of Darwin's later life was spent fighting off illness. He had caught a fever in Argentina in 1833 and was quite ill again in 1834. He suffered extreme lassitude and gastrointestinal pain, nausea, vomiting, sleeplessness and, ultimately, a fatal heart disease, all of which probably came from a parasite he picked up in South America. Though the cause of his illness was unknown at the time, today scientists think that perhaps he suffered from Chagas disease from insect bites or possibly Meniere's disease. (Chagas disease is a tropical South American disease that is transmitted by reduviid bugs; symptoms involve fever and swelling, and in its chronic form, it can have severe cardiac and gastrointestinal complications. Meniere's disease is an illness of the inner ear that is characterized by episodes of dizziness and progressive hearing loss.)

The Idea Grows

Darwin did not happen upon his theory of natural selection while on the journey; it came to him after his return to England when he had plenty of time to study the specimens he had collected and to consider how they fit together. Out of his study grew several related theories. Over time, Darwin concluded that evolution did occur, but the evolutionary change was gradual—requiring thousands to millions of years. Darwin also identified the primary mechanism for evolution: natural selection, a very controversial idea for the time (and it still is today in some parts of the Western world).

As he worked on his ideas, he began to explain it to friends as early as 1838, but he did not publish because he knew it would stir controversy. By 1844, Darwin had written a 244-

page essay explaining his early ideas, but he still circulated it only privately. In the meantime, Darwin continued working on his theory of the species. He experimented to see whether seeds could survive seawater when they were transported to isolated islands. He bred pigeons to test his ideas about natural selection being comparable to the "artificial selection" of pigeon breeders. People could relate to this aspect of his work because they understood the basics—if not the actual mechanism—of raising a faster pigeon, a better beet, or a fatter calf.

In the meantime, Darwin's reputation was growing—he had become interested in barnacles and did a great deal of writing on them, and this work earned him the Royal Society Medal in 1853, further solidifying his reputation as a biologist.

Then in the mid-1850s, another naturalist, Alfred Russel Wallace, started writing about natural selection. Charles Lyell had befriended Darwin after his trip, so when Wallace approached Lyell with the paper, Wallace contacted Darwin directly about the fact that Wallace was pursuing a similar line of thinking. Not one for a fight, Darwin offered to send Wallace's paper to a journal for him, at which point Lyell, who knew all that Darwin had accomplished, suggested that both Darwin's and Wallace's papers should be presented together to the Linnean Society in 1858. This came to pass, but Darwin was not there—his infant son had died, and he was not able to attend because of the funeral.

So why isn't Wallace ever acknowledged along with Darwin? While Wallace got started on the right foot, he was eventually ignored by the scientific community because he returned to the belief that there was a spiritual component to the process. (In a footnote to history, natural selection was also touted by a fellow named Patrick Matthew, who lived from 1790 to 1874 and wrote a book called *On Naval Timber and Arboriculture.* He extrapolated what we now think of as natural selection from his observations of how people, in raising trees of optimum quality for naval use, influenced change through artificial selection. His theory lacked scientific reasoning, but Matthew always felt wronged that he was not acknowledged for his work.)

Darwin carefully monitored the response after his and Wallace's papers were presented, and he noted that there was no real outcry. In 1859, Darwin finally published *On the Origin of Species by Means of Natural Selection, or The Preservation of Favoured Races in the Struggle for Life*, in which he established evolution by common descent as the dominant scientific explanation of diversification in nature. At the time, evolution implied creation without divine intervention, and to the greatest extent possible, Darwin avoided the use of the words *evolution* and *evolve*. The first edition of the book focused more on natural selection in nature, only indirectly alluding to it to explain the creation of humans. All 1,250 copies were sold immediately.

How Natural Selection Works

In any population, there are variations among individuals. For example, fish born with certain characteristics—strong fins and the ability to move swiftly—enjoy an advantage over other fish of the same species that lack these characteristics. If they pass these traits on to their offspring, some of their offspring that show these traits will enjoy the same advantages. Since the strong-fin organisms have a better chance to get food, they have a better chance to reproduce and pass on the strong-fin traits to their offspring. The other traits will still exist but may become less common over time. Nature is selecting for the traits that will help organisms survive in their environment. Change sometimes happens because of an alteration in the environment such as a change in climate or a change in the resources available in an area. Certain types of change may favor other qualities or traits in the fish. As the environment changes, the fish best suited to live in the new environment may fare better, live longer, and produce more offspring, until the population begins to look very different from how they originally appeared. The fish is said to be better adapted to live in the new environment. Eventually this adaptation to the change may be great enough to qualify this fish as a new species. (A species is a group of organisms capable of interbreeding in nature and producing fertile offspring.)

To Darwin, life on earth is the result of billions of years of adaptations to changing environments. Though his early work avoided mention of humans within the evolutionary process, he was no longer hesitant by the time he wrote *The Descent of Man* (1871), in which he noted that humans, like every other organism on earth, are the result of evolution.

What It Takes to Prove Evolution

In Darwin's day, he had to rely on visible observations of differences and similarities. Today we have six sources of evidence of evolution:

1. We know that all living things are made of cells and molecules. We can see the biochemical similarities between comparable compounds in different organisms.
2. Anatomy of different species can be compared, and scientists often find similar structures in very different organisms.
3. Scientists also have evidence of vestigial organs. Vestigial organs are structures found in our bodies, and the bodies of other organisms, that appear to have no current function but point to an earlier time when it would have been useful. Wisdom teeth in humans are an example of an anatomical structure that is no longer needed or useful. (Better dental hygiene has eliminated the need for these extra teeth, since we now tend to keep our teeth for a lifetime.)
4. We are now obtaining genetic evidence (DNA) that shows linkages between all living things (molecular phylogeny).
5. Embryological similarities and differences can be compared. As the embryo develops from the fertilized egg to a baby, we can see all the stages this organism went through as it was developing into its present day form.
6. The fossil record grows bigger every day. As investigative methods have improved, the indications of links among species grow in number. Scientists see that over time certain changes have occurred.

Fossil Evidence

In 1861—two years after publication of *The Origin of Species*—a significant find was sent to the British Museum of

Natural History in London. (The specimen was sold to the museum by an amateur collector, a doctor, who received the skeleton in lieu of payment from one of his patients.) In a quarry in Bavaria, a well-preserved specimen of a strange fossil with feathers from the Jurassic age was discovered and named *Archaeopteryx*. It was a confusing mix of bird and reptile that featured a birdlike head but with teeth. While Darwin had surmised that birds must have evolved from reptiles, here was the evidence, but it went unrecognized. The specimen was examined by Richard Owen, a well-respected anatomist. Thomas Henry Huxley (1825–1895), a highly regarded zoologist, paleontologist, and a great defender of Darwin's, also saw it in 1868. But neither man—nor anyone else of the day—was able to convince other scientists that this was an important piece in Darwin's theory.

Much Ado about Something

In Dayton, Tennessee, in the 1920s, a court case to test the enforcement of a Tennessee statute that prevented the teaching of evolution in public schools created a great stir. John T. Scopes, a twenty-four-year-old high school biology teacher, was the chosen vehicle for challenging the Tennessee state law that outlawed "any theory that denies the story of the Divine creation of man as taught in the Bible." The Scopes Monkey Trial, as it became known, was watched by the entire nation because of the issue and because it brought together two powerhouse legal minds: Clarence Darrow for the defense and three-time presidential candidate William Jennings Bryan as the prosecutor. The trial was held amidst a media circus. When the case went to the jury, the jurors found Scopes guilty after only an eight-minute deliberation.

A year later, the state supreme court overturned the verdict on a technicality. While it did not resolve the issue, this final step was a blow to the antievolutionists. Of the fifteen states with antievolution legislation pending in 1925, only two states (Arkansas and Mississippi) enacted laws restricting teaching of Darwin's theory.

William Jennings Bryan, Progressive politician, three-time Democratic nominee for president, and famed prosecutor during the Scopes Monkey Trial.
Source: Library of Congress.

The story is familiar to all through the stage play and the movie *Inherit the Wind.*

In 1965 Scopes wrote that as a result of the trial, "restrictive legislation on academic freedom is forever a thing of the past" (John Thomas Scopes, "Reflections—Forty Years After," www.law.umkc.edu/faculty/projects/ftrials/scopes/scopesreflections.html).

Tennessee finally repealed the law in 1967, but the story continues.

But Still a Fuss

Anyone who hears the news knows that Darwin, natural selection, and evolution are still frequently the focus of an ideological fight. Those who oppose the concept of evolution find it at odds with the literal interpretation of the origin of life as given in Genesis. The most recent anti-Darwinians are a slightly new breed. No serious biologist or geologist doubts that evolution occurred, and most supporters of intelligent design concur. They simply state that some natural processes are so complex and ingenious that they must have been created by an intelligent or supernatural being. In so stating their case, however, they are ignoring the scientific proof of natural selection that is continuing to mount because of our improved tools and testing abilities.

Currently the debates most frequently arise in the arena of school systems that are asking, "What shall we teach the children?" Some educators have been quoted as commenting that natural selection is "only a theory." However, in so dismissing it, they are ignoring the fact that in science a theory actually holds the highest rank among scientific ideas and is well supported by data and observation. They argue that there are no transitional fossils showing how one species evolved into another, but they are ignoring what has been found—or perhaps scientists just haven't found enough to fully satisfy them. Since the 1860s and the finding of *Archaeopteryx* (the bird-reptile link), paleontologists have continued to find clues. Recently they have found fossils that are transitional between whales and their terrestrial ancestors, and between finned creatures and limbed ones.

And in April of 2005, a French-led team of paleoanthropologists reported finding in central Africa a skull and other bones of a possible human ancestor that lived 7 million years ago, close to the fateful time when the human and chimpanzee lineages diverged. Scientists who studied the new fossils felt that they were a species that was "close to the last common ancestor of chimpanzees and humans" (John Noble Wilford, "Fossils of Apelike Creature Still Stir Lineage Debate," *New York Times*, April 12, 2005); other scientists wanted further proof.

The Debate Goes On

Biological evolution is the only scientific theory to have reached the Supreme Court. In 1987 the Supreme Court ruled that creationism is a religious belief that cannot be taught in the public schools. Yet discussion continues.

In 1999, the Kansas State Board of Education changed its science standards to eliminate evolution as an explanation for the development of humanity; they tossed out the big bang theory as well. The following year they reversed themselves when voters opted to oust the members who had pushed for the previous change. They have reversed again, having just approved (2005) another set of science standards that advocate exposing students to criticisms of evolution, such as intelligent design (ID).

In Dover, Pennsylvania, the right to teach intelligent design was brought to trial in late summer of 2005. Only a few weeks later, voters removed from office eight of the school board members who supported intelligent design. A few weeks after that, the judge finished his consideration of the issue and ruled against the teaching of ID in Dover.

Further Proof

Intelligent design proponents point to the human eye and the astounding ability of blood to clot to point out how difficult it is to imagine that these things "evolved." Yet scientists are at work investigating how genes work and have traced sight back to sightless bacteria from which multicelled animals evolved more than half a billion years ago. Eventually photosensitive spots gradually turned into light-sensitive craters that could detect the rough direction from which light came. Gradually the craters developed lenses, the information-gathering capacity of these organs improving all the while. (Remember the time line is billions of years.)

As for blood clotting, today' scientists are doing studies at the molecular level and are beginning to understand the order in which different proteins became involved in helping blood clot, and how eventually the sophisticated clotting mechanisms of humans and other higher animals developed. The mapping of the genome has helped with this. Scientists had predicted that more primitive animals such as fish would be missing certain blood-clotting proteins, and in fact, the sequencing of the fish genome has shown just this.

Molecular discoveries have also provided proof of shared ancestry. Now we know that every living cell makes use of nucleic acids as its genetic material and uses the same twenty amino acids as the building blocks for proteins. The development of genetics has allowed biologists to study the genetic record of evolution, and while we cannot obtain the DNA sequences of most extinct species, the degrees of similarity and difference among modern species allows geneticists to reconstruct lineages with greater accuracy. For example, when the DNA of humans and chimpanzees is compared, it is shown that the two species share 98–99 percent of our genetic material.

But There Is More to Think About

Even evolutionists aren't convinced that Darwin was totally right. In 1972 Niles Eldredge and Stephen Jay Gould in 1972 put forward the theory of punctuated equilibrium, the idea that evolution, particularly the differentiation among species, occurs relatively quickly with longer periods of little or no change. This is still under discussion.

Keep following the news. This story is far from over.

8

The Secret of Life

The ABCs (and a Little CSI) of DNA

Today you can't watch a crime show on television without seeing investigators request a cheek swab from suspects to analyze their DNA, and hardly a day passes without a news story mentioning DNA in some context. Many of these stories are crime related, but more and more, we're reading about amazing progress in science and in medicine because of the relatively new-found ability to identify DNA. And as proof of just how mainstream DNA has become, a genealogical Internet registry now offers customers the opportunity to test their own DNA and then have access to databases to see about identifying family relationships among others who have registered—amazing.

Since the late 1950s and early 1960s, molecular biologists have learned to characterize, isolate, and manipulate the molecular components of cells and organisms. These components include DNA (deoxyribonucleic acid), the repository of genetic information; RNA, a close relative of DNA, whose functions range from serving as a temporary working copy of DNA to actual structural and enzymatic functions; and proteins, the major structural and enzymatic molecules in cells.

When the structure of DNA was first discovered by two unlikely fellows, James Watson and Francis Crick, DNA was referred to as "the secret of life," but today, scientists know DNA wasn't really the secret of life. It was more like a special key—with it, they have only just begun to unlock some of the mysteries of life and development.

The Basics

Molecular biology is the study of biology at a molecular level. The field overlaps with other areas of biology and chemistry, particularly genetics and biochemistry. Molecular biology is mainly involved with understanding the interactions between the various systems of cells, including the interrelationships involved in DNA, RNA, and protein synthesis, and learning how those interactions are regulated.

Every cell within each living thing—from skin to muscle (except red blood cells) contains a copy of the same DNA. The DNA sequence is the particular side-by-side arrangements of bases along the DNA strand. The order spells out the exact instructions required to provide unique traits for each particular organism. Small structures in the cell called mitochondria (which carry out cell respiration) and chloroplasts (which carry out photosynthesis) have their own DNA and are able to self-replicate. In humans we can trace the mitochondrial DNA from mother to children and thus determine a maternal link to the children. We can do the same thing with Y chromosomes from father to sons.

Before Watson and Crick

Before DNA could even be conceived of there needed to be advances on many fronts. Ancient people had the basic idea of inheritance—they understood some basics about breeding: Certain animals, when bred together, begat animals with the same strengths (or colors, or other traits) as the parent. It was somewhat of a hit-or-miss process, however, and no one had any clue about the mechanics of how this worked scientifically.

Before genetics could become a science, the world needed to discover how life began (scientifically), and they needed to know that there were mechanisms—genes—in our bodies that somewhat reliably determined inheritance.

The microscope work of Antony van Leeuwenhoek (1632–1723, introduced in chapter 6) led to one of the first clues that was needed. Not only was Leeuwenhoek the first person to see sperm under a microscope, but Leeuwenhoek's discovery of flea eggs dashed the theory that fleas—or any living thing—could grow spontaneously.

Then in the mid-nineteenth century, an Austrian monk, Gregor Mendel (1822–1884), pursued a hobby of growing garden pea plants that he crossbred. Mendel kept careful records of his crossbreeding of tall or short plants that had smooth or wrinkled peas, and he began to see that something within the plant-breeding process operated with an orderly, dominant/recessive plan. (When he bred tall plants with short ones, he didn't get medium-sized ones as one might expect; he always got tall ones.) There was not much of an avenue for a monk with an interesting hobby to tell others what he had found, so no one knew of his contributions during his lifetime. But he left behind meticulous records, and eventually his work was brought to light and was to lay the groundwork for some very significant discoveries.

It wasn't until 1900—the date of the onset of modern genetics—that an unappreciated and almost forgotten paper written in 1865 by Mendel (who by 1900 was no longer living) began to send reverberations through the scientific community when it was introduced by scientists Hugo deVries and Carl Correns. At that time, scientists were beginning to grapple with the meaning of inherited traits.

Though the "so *that's* how it works" moment for DNA didn't occur until the 1950s, the existence of DNA was discovered in the early twentieth century by a German biochemist who found that nucleic acids were made up of sugar, phosphoric acid, and several nitrogen-containing compounds known as bases.

Not until 1944 did American scientist and early molecular biologist Oswald Avery (1877–1955) prove that DNA carries genetic information. While this discovery fascinated scientists, there was still little they could do with the information because they did not yet know the molecular structure of DNA. Only by decoding it could they begin to understand it—and to understand how to use that information to help humankind.

Scientists and Serendipity

While almost all scientific discoveries rely heavily on the scientific work that preceded it, the discovery of the structure of DNA relied not only on the work of others, but also a great deal on serendipity. And like a good horse race, there were several scientists (including teams of scientists) making the run down the final stretch to successfully identify the structure of DNA. Watson and Crick's victory was due to a little luck and a lot of careful attention paid to their own work as well as what they were hearing from others. Added to this mix was the fact that one of the front-runner scientists (Rosalind Franklin) was abrasive and therefore not well liked, which encouraged members of her department to share with others some of what she was finding.

The story began at King's College in London, where British biophysicist Maurice Wilkins (1916–2004) was at work to determine the structure of the DNA molecule. Wilkins decided that to understand DNA, he needed an image of it, so he turned to Rosalind Franklin (1920–1958), who was only a graduate student but was acknowledged to be the best person at X-ray crystallography. Franklin set to work, and though she did not like to provide interim reports, she agreed to give a departmental talk about her early images of a dry and a wet form of DNA, which were beginning to provide evidence of the helical structure of DNA (1951). As it happened, James Dewey Watson (1928–), a twenty-three-year-old American zoology graduate who had studied ornithology and then viruses, visited the department that day and gained some important clues about Franklin's early findings.

Watson had only recently arrived at Cavendish Laboratory, where he hoped to study DNA. He had just become interested in the subject, but he had already decided that he needed to be the one who decoded the structure. It was at Cavendish that he encountered his future partner, Francis Crick (1916–2004), a physicist who had become fascinated by the application of physics to biology and who was also intent on investigating how genetic information might be stored in molecular form.

Though an odd pair, Watson and Crick were united by their determination to better understand DNA. Crick brought his knowledge of X-ray diffraction, and Watson brought knowledge of viruses and bacterial genetics. Ironically, Watson and Crick had been told by their superior that they were not to work on the unraveling of DNA, as other scientists were working on that. Despite this, Watson and Crick continued—going as far as to invite Franklin to meet with them about their work. She dismissed them, however, and she refused to come.

On another continent, American Nobel Prize–winning chemist Linus Pauling (1901–1994) was working with X-ray crystallography and molecular model building. In 1952 he hoped to travel to England to a conference he knew Franklin would be attending, but he was denied a passport because of allegations that he was a Communist sympathizer. He was far enough along in his work that had he been able to come, he and Franklin almost certainly would have been the first to come up with the solution to the structure of DNA.

In the meantime, Franklin had made great progress with her work. She had successfully developed a good photograph of the B (or wet) form of DNA, which showed a double helix. However, she wasn't ready to release her information until she had further explored something that was bothering her about the A (or dry) form. Her reluctance annoyed her partner, Wilkins, who decided to move on without her.

In 1953, Watson dropped in to visit Wilkins, and since both of them had suffered being dismissed by Franklin, they bonded over their annoyance at her attitude. During the course of their meeting, Wilkins happened to show Watson a copy of Franklin's photograph of the wet form of DNA, revealing the helical form that Watson suspected. The photograph further lead Watson to suspect that DNA could reproduce because it was structured as a double helix.

Crick was aware of the tension between Wilkins and Franklin and used this as an opportunity to approach his superiors for permission to make another attempt at a model of DNA; this time permission was granted.

Later on, without Franklin's permission, a departmental report of Franklin's was passed to Watson and Crick. It offered conclusive evidence that DNA was a multiple helix, and Watson and Crick learned that the phosphate backbones of DNA should be on the outside of the molecule. This finding was key to figuring out the structure. (It will long be debated whether Watson and Crick should have had access to Franklin's results before she formally published them herself.)

For the final piece of the puzzle, Watson and Crick turned to work done by Erwin Chargaff in 1950. Once again, Franklin was the scientist who really had a full understanding of Chargaff's work, which involved the base pairings of DNA. She had completed a draft of a paper, dated March 17, 1953, that identified the double-helix structure of DNA as well as the specific base pairings that permit the "unzipping" of the double helix during cell division so that the gene, and eventually the chromosome, can replicate.

Despite Franklin's overall understanding of the process, it was Watson and Crick who got to press first with their paper. In a rather low-key way, Watson and Crick's paper on the structure of DNA appeared in the British journal *Nature*, and it described the DNA molecule as a long, two-stranded chain coiled into a double helix and resembling a twisted ladder. Their paper summed up the contribution of Wilkins and Franklin by simply mentioning that Watson and Crick's thoughts were "stimulated" by the unpublished results of Wilkins, Franklin, and their coworkers at King's College.

In 1962, Watson, Crick, and Maurice Wilkins were given the Nobel Prize in Physiology or Medicine for their work. Franklin had died in 1958 from cancer, possibly related to her extensive exposure to radiation in her work with the X-ray diffraction that was so vital to better understanding the structure of DNA. Because the Nobel cannot be given posthumously, she was never honored for her work, so it is left to the retelling of the story to acknowledge her very considerable contributions.

The Human Genome: What It Is

The human genome is a person's complete set of DNA, arranged into 23 distinct chromosome pairs (the 24th pair is the set that determines gender)—physically separate molecules that range in length from about 50 million to 250 million base pairs. Each chromosome contains many genes, the basic physical and functional units of heredity.

Genes make up only about 2 percent of the human genome; the remainder consists of noncoding regions, whose functions may include providing chromosomal structural integrity and regulating where, when, and in what quantity proteins are made. Although genes get a lot of attention, it's the proteins they make that perform most life functions and even make up the majority of cellular structure.

The Story Continues

By the 1980s, James Watson had another passion to pursue, and this, too, has been absolutely key to medical and scientific advances. Watson helped lobby Congress to create the U.S. Human Genome Project, the multimillion-dollar effort to map out the exact nucleotide sequence contained in each of the 24 human chromosomes—the so-called book of life (consisting of approximately 3 billion letters).

Ironically, the Human Genome Project grew out of the U.S. Department of Energy (albeit, it did originate in its Health and Environmental Program). Since 1947 the DOE and its predecessor agencies have been charged by Congress with developing new energy resources and pursuing a deeper understanding of potential health and environmental risks posed by their production and use.

In 1986 Charles DeLisi, who was then director of DOE's health-related research programs, became convinced that if they were going to be able to effectively study the biological effects of radiation (along with information on whether these effects were passed on genetically, such as in the cases of survivors of Hiroshima), he needed a way to do so quickly. By 1990, DOE and the National Institutes of Health, who understood that knowledge of the human genome was necessary to the continuing progress of medicine and other health sciences, had agreed to devote $3 billion to the project, and they established a fifteen-year timeline. They were joined by the Wellcome Trust, a private charitable organization in the United Kingdom. There were also contributions from Japan, France, Germany, and China.

In 1998, a private firm, Celera Genomics, run by researcher Craig Venter, entered the picture. Venter was using a newer technique (shotgun sequencing, a process that breaks the DNA into shorter segments that permits faster readings; the segments are rejoined for a complete reading at the end). He hoped to finish before the government and planned to patent some of what he found. Though his contributions were notable, in March 2000 President Clinton announced that the genome sequence should not be patented, sending biotech stocks—including Celera's—plummeting.

The competition proved healthy, and by 2000, due to widespread cooperation, both public and private, a rough draft of the genome was actually finished early and was jointly announced by then-president Clinton and British Prime Minister Tony Blair.

In February of 2001, both Celera and the government scientists published details of their discoveries—*Nature* published the government's version, and *Science* published Celera's. Together, the sequence they had identified made up about 90 percent of the genome. By 2003 a joint release announced that 99 percent of the genome has been sequenced with 99.99 percent accuracy. For all ostensible purposes, the project was completed in April 2003, bringing it in two years ahead of schedule. This happened to be the fiftieth anniversary of Watson and Crick's publication of DNA's structure that launched the era of molecular biology.

What has surprised everyone—scientists included—is that humans are a great deal "simpler" than what they had originally imagined: They've found that the human genome has only about 30,000 genes; the original estimate had been three times that number. This finding suggests that a great deal more has to be learned about how genes function and how their instructions are carried out as well as how they produce diseases and other anomalies. But in the meantime, the gains have been incredible.

DNA and the Future

Knowledge about DNA may lead to understand how all the parts of cells—genes, proteins, and many other molecules—work together to create complex living organisms. DNA underlies almost every aspect of human health, and understanding what DNA has to do with health will have a profound impact on the ways disorders are diagnosed, treated, and prevented. For example, scientists have discovered a variant gene carried by more than a third of the American population that leads to a sizable increase in the risk of Type 2 diabetes. This should lead to an improvement in diagnostic testing as well as in treatment.

In December of 2005, a new project, the Cancer Genome Atlas, was announced. The Cancer Genome Atlas Pilot Project is designed to identify and unlock the genetic abnormalities that contribute to cancer—an effort that could lead to new diagnostic tests and treatments for the disease. Scientists have long known that genetic mutations accumulate in a person's normal cells over a lifetime and can make those cells cancerous. About 300 genes involved in cancer are already known, and there are a handful of drugs that work by interfering with specific genetic abnormalities.

New accomplishments are announced almost every week. In 2005, the sequencing of the dog genome was completed, taking its place with the completed sequencing of mice and chimps. All of this study will lead to deeper insights into evolutionary history and a greater understanding of more and more species.

All this information about the various genomes is now available to researchers worldwide, and a brand new future is opening for the human genome reference sequence. Among the changes that you may see are these:

- Exploration of the evolutionary process: Scientists are comparing the lineage of chromosomes in various species to better understand what changes there have been in various organisms and why they occurred.
- Gene testing: Already companies are beginning to offer inexpensive and easy-to-administer genetic tests that can show a predisposition to a variety of illnesses such as breast cancer, blood clotting, cystic fibrosis, liver diseases, and so on.
- Gene therapy enhancement: One day one's own tissues could help replace cells damaged by injury or disease. DNA could be withdrawn and used to prime embryonic cells that are ready to serve as replacement cells to bolster normal function or to create immunity.
- Pharmacogenomics: We will see a movement away from one-size-fits-all medicine. In the future we will see both treatments and vaccines that will be made to order for each individual. This will offer direct life-saving benefits, as today more than 100,000 people die each year from bad reactions to medications. DNA will help predict what will work for whom.
- Improved vaccines: In 2006, the government announced a possible avian flu vaccine, and scientists think they will be able to identify the exact flu strain and manufacture it quickly with the help of genetic engineering.
- Genetically modified foods: Scientists are already beginning to create healthier crops based on understanding the genetics of different plants.

Ethical Issues Will Need to Be Solved

Right alongside the scientists making all the genetic advances, scientific committees will need to wrestle with the ethical issues that these advances raise. These can range from the seemingly simple (what happens if genetically altered corn cross-pollinates with regular corn?) to issues that will affect our daily lives in larger ways. For example, if someone undergoes genetic testing that shows a predisposition to some type of chronic illness, we have to be certain that these issues can never be raised or used to affect what happens with that person's job. Predisposition to a disease does not mean you will get the disease. It means that if the correct influences occur, you have a good chance of getting the disease. For example, if you have a predisposition to lung cancer then being exposed to secondhand smoke or pollutants in the air could cause certain cells to change and develop into lung cancer, sparking the nature versus nurture debate (nature—the DNA you inherited—versus nurture—the environmental factors that could activate your DNA).

Questions about privacy, fairness in use, and access to genetic information will all need to be answered as advances continue.

Today you have to read your newspaper or the appropriate magazines and journals—not your science book—to keep up with what is happening with DNA.

DNA and the Criminal Justice System

Given DNA's use in television crime shows and its capturing of so many headlines, a chapter on DNA cannot end without talking about its use in the criminal justice system. Though the sequencing of DNA has now been around since the 1950s, its use by the criminal justice system is a relatively new development.

DNA was first used in a criminal case in Britain in 1986 when Professor Alec Jeffreys assisted in solving a pair of rape-murders of two teenagers. (Not until 1992 was the use of DNA officially approved for court cases by the National Academy of Sciences.) When two fifteen-year-old girls were murdered in 1983 and 1986, police originally arrested a young man with a history of mental illness, but Professor Jeffreys wasn't convinced that the police had the right man. After analyzing semen collected from both bodies, Jeffreys asked everyone in the small town of Narborough to voluntarily submit to DNA testing. At first, no guilty party emerged. Eventually police found that a baker by the name of Colin Pitchfork had

Imagine What Else We May Learn

In January of 2006, Japanese scientists announced that they had located the gene that dictates what kind of earwax people have. It turns out that earwax comes in two types: wet and dry. The wet form predominates (97 percent of the population) in Africa, and people in Europe have the wet kind. The dry kind is found more frequently among eastern Asians. People in southern and central Asia are roughly half and half.

While you may not care that much about earwax, perhaps this additional piece of information will interest you: The scientists write that earwax type and armpit odor are correlated. Populations with dry earwax, such as those of eastern Asia, tend to sweat less and have little or no body odor. Those with wet earwax tend to sweat more and have more body odor.

---

paid someone else to give a blood sample for him. When police reinterviewed Pitchfork, he confessed to both crimes. And when he provided his own DNA, the police had a match and their man.

An early test of DNA in the courtroom in this country occurred in what has become known as the "trial of the century"—as the 1995 criminal murder trial of sports star O. J. Simpson was often called. Though Simpson's DNA was found at the crime scene, the prosecution was unable to convince the jury of Simpson's guilt. The use of DNA in court was still very new, and though scientists and lawyers faced off on the issue, the defense successfully convinced the jury that the evidence may have been contaminated. This case brought home the importance of careful training of criminal investigators in the collection of evidence, and also as a result, crime laboratories realized the importance of being certified for this type of testing.

By the late 1990s, forensics labs started to adopt a new method of analysis called STR (short tandem repeats) that cuts analysis time from weeks to days and uses patterns that repeat just a few times, between five and thirty in most cases. And it also improves accuracy—early on in DNA forensics, the chance of error was one in one hundred thousand. The STR method makes it more like one in a trillion.

Use of Fingerprints

As DNA gradually edges out the use of fingerprints in criminal investigations, it is interesting to look back at the first time fingerprints were used in a murder case. With this type of evidence, too, it took thirty years of experimenting before it became accepted.

In 1905, two shopkeepers, a husband and wife, were found bludgeoned to death in their shop in Deptford, a small town outside London, and their cash drawer had been robbed of ten pounds. An Inspector McNaughton, who had been learning about and experimenting with fingerprints, was among those called to the scene. When he examined the cash box, he found on the underside of it one sweaty thumbprint. He sent it off to be identified, and this evidence soon led to the arrest of one of the two brothers who were responsible for the crime.

A major advantage of DNA over fingerprints at this point is that DNA evidence has been saved. A computer database of DNA from people who have had encounters with the law has permitted the police to revisit cases. Many prisoners have been freed based on the fact that their DNA did not match that left at a crime scene, and in many cases, criminals whose DNA has popped up in the system have been connected to long-ago murders and rapes.

9

The Avian Flu

A Dangerous Virus and Its Long Shadow

At this writing, the avian flu story has the potential to be the biggest science story of the twenty-first century. Scientists are quite concerned that this virus that is spreading through the world's bird population could mutate to spread among the human population, becoming a worldwide pandemic.

Past pandemics have spread globally in two and sometimes three waves, and as the world has become more interdependent, it is guaranteed that any major epidemic of the next few years will be a global one. If this one were to spread, it is estimated that 20 percent of the world population would become ill; 30 million people would need to be hospitalized, and a quarter of those people would die. The U.S. Congressional Budget Office has estimated that a severe pandemic could infect 90 million Americans, kill 2 million, and push the economy into a recession. Schools would close; absenteeism at workplaces would be high, and air travel would fall by two-thirds. In October of 2005 President Bush announced that he would consider using the military to enforce quarantines in the event of an outbreak in the United States. (To understand the possible collapse of our community infrastructures under these circumstances, we need only think back to September of 2005 in New Orleans, where the world witnessed the collapse of every type of support system that should have helped Gulf area residents through the storm emergency.)

Even in a scenario that involves a more mild form of the influenza, the Congressional Budget Office predicts a drop in demand for all industries of about 3 percent, a 17 percent decline in travel, and a 20 percent drop in recreation and restaurant attendance.

The scientific community has been warning of this danger for several years, but only recently have the experts been able to capture the governmental attention they need—and that the issue deserves. The embarrassment over the emergency response to Hurricane Katrina, added to the fact that within the bird population this new avian flu is definitely on the move, may partly explain the awakening of the United States to the threat. Other countries may be paying attention because, as wild birds carry the disease along the bird flyways, they are seeing their domestic poultry and ducks decimated by the disease.

So how desperate is the situation? As the chapter explains, no one is quite sure. Global influenza pandemics tend to occur about every thirty years, and to qualify as a pandemic, three conditions must be met:

1. A new virus subtype emerges.
2. It infects humans, causing serious illness among a large proportion of the population across a wide geographic area.
3. It spreads easily and sustainably among humans.

At this writing, the first two conditions have already been met, but perhaps we'll be lucky and the virus will never mutate into something that is transmitted among humans. But even if we are that lucky, this flu needs to be viewed as a fire bell in the night. As you see in this chapter, we desperately need a global disease surveillance system that watches for illnesses in both humans and animals to protect humankind.

So Why the Big Worry about This Flu?

The H5N1 influenza virus is a Type A deadly strain. (The name is an abbreviation for a scientific description of its structure.) The disease has actually been wiping out chickens and ducks, primarily in Asia, for a good number of years. The first transmission to a human occurred in Hong Kong in 1997, and to date, the disease appears to kill about half of the people who are infected. In comparison, the seasonal flu that strikes each

The Terms and Statistics

SEASONAL FLU: A contagious respiratory illness that is caused by an influenza virus and often occurs in the winter.

EPIDEMIC: An outbreak of a disease that affects many people at the same time in a community.

PANDEMIC FLU: An epidemic that reaches global proportions. There were three pandemic flu outbreaks in the twentieth century: in 1918, 1957, and 1968. There are three conditions for a pandemic: A new virus emerges; it infects and causes serious illness to humans; it spreads easily among people.

BIRD FLU: An illness caused by an avian influenza virus, which occurs naturally among birds and in rare cases is transmitted to humans.

H5N1 FLU: A strain of bird flu that is highly contagious and rapidly fatal when it has spread from birds to humans. The H and N stand for proteins that cover the virus.

STATISTICS ON FLU DEATHS FROM PANDEMICS

1918 SPANISH INFLUENZA: 50 million worldwide. (This flu killed more people than World War I. Unlike the case with most flu epidemics, most of the casualties were otherwise healthy people ages fifteen to thirty-four.)

1957 ASIAN FLU: 700,000 in the United States.

1968 HONG KONG FLU: 500,000 in the United States.

ANNUAL SEASONAL FLU: 400,000 in the United States.

Source: Centers for Disease Control and Prevention

---

winter kills only a small fraction of those who become ill—that's still about 36,000 Americans.

Thus far, the virus is being spread via migratory birds that seem to be carriers, so they can fly for a few hundred miles after infection. Then for a week or so they seem to shed the virus into the lakes or marshes where they land. Other wild birds become infected, fly away, and further spread the disease. Domestic fowl, who encounter the wild birds or the water where they have been, become ill. In the domestic birds, the virus seems to be more virulent and kills quickly; it is thought that the virus becomes more virulent in populations that live in close quarters.

Until spring of 2005 the virus was contained to Southeast Asia. By July of 2005 infected birds had carried it to Siberia, where the north-south flyway meets the east-west flyway. As a result it has now spread as far as Turkey and Iraq.

Since 2003, the World Health Organization has reported that patients with the H5N1 virus have been identified in Vietnam, Cambodia, China, Indonesia, Thailand, and, most recently, Turkey and Iraq, with the number of people infected and dying increasing each year. Thus far, there have been about 140 human cases in Asia, according to the World Health Organization, and, most recently, at least 15 in Turkey. The official numbers of cases and deaths includes only those confirmed by the World Health Organization, making it possible that others have sickened and died without seeking medical treatment.

In January 2006, it was announced that a fifteen-year-old Iraqi girl died of bird flu after touching a dead bird infected with the disease. Because of Iraq's close proximity to neighboring Turkey, no one was surprised that the disease had spread; however, flu experts note that this is their nightmare—that it enters the human population before anything can be done. If the outbreaks are reported when it is still in the bird population, there are ways to try to stem the spread of the disease. Once it has infected an area for long enough for it to spread to people, it is a great deal harder to control. For example, Turkey allowed the disease to travel throughout the country, so it is now trying to contain 55 outbreaks in fifteen provinces.

To date, it appears that all those who have caught the deadly disease were in direct contact with infected ducks, chickens, or possibly pigeons. There is not yet any definite evidence that the virus is being transmitted person to person. But viruses mutate rapidly, and the fear is that the virus will evolve into a form that can be spread among people.

There is a "bad news, good news" aspect to this bug. The bad news is that thus far the virus has killed more than half the people it has infected (as a result of bird-to-human transmission). If there is good news—or hopeful news anyway—it's that H5N1 has circulated in Asia for about a dozen years and has not yet crossed over to person-to-person transmission.

How the Virus Can Mutate

Scientists know that there are two possible ways for the virus to become one that can be transmitted between people. The first occurs when avian and human flu strains combine genes. This genetic mixing can occur when a person is infected by both avian flu and a human flu strain at the same time.

The second possibility is that the virus itself may undergo enough mutation to make the leap. A recent analysis of the 1918 flu strain (see pages 109–110) indicates that it may not take much for the flu to mutate on its own.

Is This Another "Boy Who Cried Wolf"?

The scenario painted by the experts is only speculation—albeit educated speculation—that most world leaders are taking seriously. And as always, there is collective memory of "the boy who cried wolf."

In 1976 there was great concern about a swine flu that was beginning to infect people, and under Gerald Ford the U.S. government tried to gear up quickly for a nationwide vaccine program. Eventually one-third of all adults were vaccinated, but the swine flu epidemic never came to pass. In the final analysis, more people suffered from the vaccine's side effects—including the debilitating Guillain-Barre syndrome—than ever suffered greatly from the swine flu.

This makes governments gun-shy. The difficulty is assessing the risk and deciding if the avian flu is a real threat. And actually, as of 2006, the world is assuming it is all too real a possibility.

What Needs to Be Done

The ideal plan for dealing with H5N1 influenza was to step in to control the disease before it got established. This would have involved eliminating the virus in domestic poultry overseas (vaccinate the birds or eliminate them before the virus spread to migratory birds) before it could mutate into a strain that passes among people. Early detection and rapid response to bird flu on a global scale could drastically cut the costs of dealing with a full-blown human pandemic.

Since it seems that we have already lost the ability to control the virus in this way, we must go on to Plan B, which begins with planning for mutations that allow the virus to be transmitted between humans. The two avenues to fight the virus are vaccinations (the better option) and antiviral medications.

Creating a vaccine isn't easy. Each year, the makers of seasonal flu vaccines have to take an educated guess as to exactly which flu they think will be circling the globe the following winter. Existing vaccines are powerless against the H5N1 strain, so to combat it with a vaccine will require trying to anticipate any mutations of the virus so that any vaccine created will fight the right bug.

The vaccine-manufacturing process is in itself an ordeal of sorts. Up until recently, the process involved injecting a flu virus into a fertilized chicken egg, mixing a "wild" virus with a "lab" virus to assure that the virus would replicate within the egg so that it could become a vaccine. The "egg method" takes nine months from when a virus is identified as a risk to the distribution of a vaccine, and this certainly does not provide enough time for a world supply to be created.

Now scientists have a more rapid process for creating vaccines that employs reverse genetics. With this method, the live flu virus is stitched into loops of DNA called *plasmids.* The plasmids assemble into whole flu viruses in the lab. By using reverse genetics, scientists can create exactly the types of virus they need without having to mix viruses, such as adding the lab virus. They should be able to create the basic vaccine in a much shorter length of time (possibly as quickly as four weeks).

If there is not enough time to create a vaccine—or if the one created is ineffective—then doctors would turn to the antiviral medications. Two drugs, oseltamivir (commonly known as Tamiflu) and zanamivir (Relenza) may reduce the severity and duration of the illness. Both are from the neuraminidase inhibitors class (they inhibit the activity of the flu virus protein), and they must be started within forty-eight hours of symptom onset. Often antiviral medications become ineffective if overused, so there is no guarantee that they will work when they are needed.

What They Have Learned from the 1918 Flu Virus

The H5N1 influenza virus is closer in composition to the Spanish flu of 1918 than it is to other viruses that have circulated, so for that reason, the 1918 virus has been of interest.

The Spanish flu seems to have begun as a normal human flu strain that circulated in the United States in the spring of 1918. By summer the virus had reached the battlefields of World War I. There, among people packed into the trenches, trucks, trains, and hospitals of the western front, it turned lethal, just as the avian flu does among birds on crowded chicken farms today.

Because scientists of that day had not learned to isolate viruses for study, there had been no data and no simple way to learn any lessons from that pandemic. Dr. Jeffery Taubenberger, chief of the molecular pathology department at the Armed Forces Institute of Pathology in Washington, felt that the Spanish flu had stories to tell, so he developed a plan to recreate the virus to study it. In a ten-year project completed in the fall of 2005, Taubenberger and his team isolated and began to study the Spanish flu.

Working under extremely secure conditions so as not to endanger the staff or permit the escape of the virus, Taubenberger and his research group located lung tissue from two soldiers who died of the 1918 flu from an autopsy tissue warehouse established by President Lincoln. The importance of careful handling of the virus was brought home very recently when some testing on the SARS (severe acute respiratory syndrome) virus in Beijing resulted in several laboratory workers becoming infected with the virus from working with it.

A third sample was sent to him by a retired pathologist, Johan Hultin of San Francisco, who spent his own money to go to Alaska, where he gained permission to excavate a body from a mass grave in a community where 72 adults died from the flu in 1918. Because the grave was in permafrost ground, the bodies had remained frozen all this time, and he snipped frozen lung tissue from a woman and sent it off to Taubenberger for his study. Taubenberger himself was busy procuring samples from the Smithsonian of birds that had died in 1918–1919 so that he could identify for certain that this was a bird virus.

Though it took a full decade to piece together the virus, researchers have not only analyzed the virus, but they have also conducted gene-swapping experiments to determine what weakens the virus and what makes it stronger.

In comparing the 1918 flu with today's human viruses, Taubenberger found that the virus had alterations in just 25 to 30 of its amino acids. Those few changes—much fewer than expected—turned a bird virus into a killer virus that could spread from person to person. The current avian bug has already made 5 of the 10 changes found in the 1918 virus.

Been There, Done That

While the possibility of an avian flu pandemic has the attention of both the press and the public, it is interesting to look back and see that other illnesses have also caused this same type of alarm. On July 13, 1883, the *New York Times* ran a notice about the spread of cholera in Egypt, and that article was accompanied by stories from both Paris and London. In Paris, the Hygiene Commission agreed that Louis Pasteur should journey to Egypt to offer help; in London, Mr. Gladstone announced to the House of Commons that the government would send a British Surgeon-General who had experience in treating cholera.

The Future Involves Zoonotic Diseases

In the last thirty years, there has been a big increase in the number of new diseases that have jumped from wildlife to humans, and these have proven to be quite harmful to humans; AIDS, Ebola, Lyme disease, avian flu, West Nile virus, monkeypox, and SARS are among them.

Because of rapidly changing human behavior and animal

ecology, infections are spreading farther and faster. When multiple species of animals come together at high density, it becomes possible for pathogens to jump between species, and today more than half of emerging diseases have been transmitted from animals to people. These diseases are permitted to emerge because of changes in human activities such as wildlife trade and global travel.

Why We Need to Pay Attention

Disease knows no borders. While localized outbreaks might have occurred in years past, today an outbreak in one country can very rapidly become a problem for countries on the other side of the world. Even if the threat of avian flu passes and leaves us relatively unscathed, the world must not become complacent.

We need only think back to 2003 and the SARS (severe acute respiratory syndrome) outbreak. For several months SARS held the headlines of every major news vehicle. This disease quickly spread to 30 countries in nearly as many days claiming more than 800 lives and terrifying millions of people. Fear gripped the United States when an outbreak occurred in Toronto, proving that the illness could move out of Asia and go almost anywhere. We were all spooked by the news coverage of the citizens of Hong Kong and the major cities in eastern China who wore face masks everywhere to try to prevent transmission of the illness. Travel restrictions were put in place, and the disease crippled tourism and airports in several countries and caused many other ripple effects from school closures to quarantines to various forms of discrimination.

The slowing of SARS began when scientists traced the source to a coronavirus that was harbored by civets and raccoon dogs, considered culinary delicacies in southeastern China, but ultimately traced back to bats as the ultimate source. Marketplaces that featured wildlife trade—cages with tightly packed populations of animals for sale—were quickly closed down, and the spread of the disease began to slow.

China's reluctance to come forward to admit the outbreak

was largely responsible for what eventually was a temporary crisis. Had the government been willing to pay attention to what was happening earlier, perhaps the fear and the travel bans would never have had to go into effect.

Both SARS and avian flu originate in wildlife reservoirs (organisms that host parasites that are pathogenic for some species but do not damage the hosts), and the best way to slow the illness is to remove the reservoirs. It's crucial to identify the reservoirs of these emerging pathogens so that scientists can understand how the diseases emerge and predict and prevent future outbreaks.

Hope for the Future: Conservation Medicine

Scientific study has revealed that damaged ecosystems—characterized by toxins, degradation of habitat, removal of species, and climate change—have created conditions for pathogens to move in unexpected ways, including from animals to people. The Wildlife Trust created the Consortium for Conservation Medicine, with the intent of combining efforts among scientists and medical personnel to interrupt transmission routes and prevent future disease outbreaks.

Close to home, scientists point to what has happened with Lyme disease. While the pathogen that causes Lyme disease has been around for a long time, it didn't create a problem for humans until recently as we chopped down forests to make room for suburban homes.

Lyme disease is carried to humans primarily via ticks that often travel on white-footed mice, who get the disease from the ticks that suck their blood and are perfectly happy living in the suburbs. What is missing are the animals who would have lived in the forests and would have helped keep the balance of nature. Chipmunks, weasels, and foxes are all poor reservoirs for Lyme disease, so they could live in coexistence with the ticks without spreading it, and they also prey on white-footed mice, reducing their numbers. As a result, a tick in the Adirondacks is less likely to carry Lyme disease than a tick in the suburbs.

A similar interference with nature has happened as an increasing amount of the Peruvian rain forest has been destroyed. There has been an explosion of malaria-bearing mosquitoes that thrive in the sunlit ponds created by logging operations. (For more information about malaria, see chapter 5.)

This is a global issue that requires worldwide attention to both the ecology of our planet and what's going on in both the human and wildlife populations. If we pay attention and act, we might make a difference.

PART THREE

Amazing Discoveries That Changed Our View of the Universe

10

The Copernican Revolution

A Four-Hundred-Year Story

When we think of a revolution, we tend to think of a defining moment when everything changes. When you really examine a revolution, however, you see that it actually tends to be slow in coming and involves a complex reordering of thought. What we refer to as the Copernican revolution is a perfect example. Though Copernicus did most of his work in the early 1500s, this revolution of his actually stretches back to the second century BC, hops forward to the sixteenth century, where it picks up speed, and is finally "put to rest" in 1992, the twentieth century.

As for the revolution—what a tale it is. It involves Copernicus himself, of course, as well as several other famous scientists, a philosopher, the Catholic Church, and the sun and all the planets that were known in the sixteenth century. Deception, politics, and punishment of two of Copernicus's disciples—one was placed under house arrest and the other was burned at the stake (though to be honest, his Copernican philosophy was only part of the cause)—are all part of this amazing story.

But to understand why Copernicus is credited with a revolution, we have to take a step back in history to understand people's perception of the universe. Though they had no instruments to view the heavens, early Greeks had identified several of the planets and knew they traversed the sky (the word *planet* is from the Greek word for "wanderer"). To explain the inexplicable, people created myths, so the Greeks explained the movement of the sun by telling of the Greek god Helios, who would wake to the crowing of the roosters and ride in a chariot of fire pulled by four horses across the sky each day. When he reached his palace in the west, he used a golden ferry boat to cross the ocean back to his eastern palace. The next day he repeated the trip.

Myth was eventually replaced by science. The study of the heavens, soon to be known as astronomy, was one of the first sciences to develop. People were beginning to realize that the movements in the heavens were predictable—and therefore, they could be studied. They also realized that observations of the phases of the moon could be recorded—and predicted—with a calendar.

By the second century AD, Claudius Ptolemy, an Egyptian-born Greek astronomer and geographer, put forth the thought that the heavens were a series of rotating spheres that contained and moved the planets and stars. He believed the planets and sun orbited Earth in the order Mercury, Venus, sun, Mars, Jupiter, Saturn. His system predicted planetary motions, eclipses, and many other heavenly happenings, and for almost 1,500 years it was the best explanation that anyone had for what was happening in the universe as they knew it.

The Calendar We Use Today

In 46 BC, Julius Caesar introduced the Julian calendar (with 365 days and an extra day added every fourth—or leap—year), and it is the basis of the one we use today.

For generations the main task of astronomers was to develop and keep tabs on the calendar, a vital element for agrarian success as well as success on the battlefield, and important, too, for the appropriate timing of religious observances.

Emerging from the Middle Ages

At the time of Copernicus (1473–1543), Europe was just emerging from a period when there had been few new developments in science, art, or literature. However, by the sixteenth century, the Catholic Reformation was underway. Scholarship was again beginning to be valued, and intellectuals were taking another look at classical writings.

Copernicus's uncle, a bishop, arranged for him to be made canon of Frombork (Frauenburg), Poland, a post that he retained for life. Because the duties were light but the income was adequate, Copernicus could pursue his interest in astronomy. While Ptolemy's complex mathematics that explained so much about the heavens was still held as the paradigm, astronomers, including Copernicus, were beginning to reevaluate some aspects of the Greek astronomer's work. There were observations that didn't fit with what they were seeing and some mathematical calculations that didn't work out. Copernicus also became intrigued with a theory that Greek astronomer Aristarchus (ca. 310–230 BC) had espoused—that the universe did not revolve around a stationary earth; he suggested that it was the sun that was the center of the universe. Copernicus began to work with the idea that the planets—including the earth—orbited around the sun. By 1513 Copernicus was ready to write and share with others a brief outline of his new ideas.

Over time, Copernicus formalized his thinking in *On the Revolutions of the Heavenly Spheres.* In the first section of the book, Copernicus introduced his belief that the universe was heliocentric (the sun was the center). The remaining 95 percent of the book was devoted to mathematical formulas supporting the hypothesis. Though many of his geometric models were still based largely on Ptolemy, Copernicus wanted to further explore whether the sequential distances of the planets as well as their orbital periods could be calculated more precisely with a sun-centered rather than an earth-centered perspective.

Perhaps it was his feeling of duty to the Church that had provided the opportunity for his studies, or maybe it was out of concern for being branded a lunatic or a heretic, but Copernicus did not look to publish his work. Certainly, this decision might have been affected by the fact that information in those days was spread primarily by word of mouth. Few people of the day could read—and since scholarly work of that time was written in Latin, the audience for written material was very limited.

However, others felt it was important that Copernicus be published. George Joachim Rheticus (1514–1574), a teacher

The Copernicus memorial in Warsaw, Poland.
Source: Library of Congress.

of mathematics at the University of Wittenberg, had become acquainted with Copernicus in 1539, and Copernicus had permitted him to publish his own *First Report* on Copernicus's work. Perhaps because no major controversy erupted over Rheticus's work, Copernicus permitted Rheticus to take the complete manuscript of *On the Revolutions of the Heavenly Spheres* to Nuremberg, the nearest location where a book could be printed.

An Introduction Is Added Secretly

Rheticus's teaching duties kept him from staying to supervise, and he asked Lutheran clergyman Andreas Osiander (1498–1552) to take charge of the printing. Unbeknownst to Rheticus and Copernicus, Osiander inserted his own unsigned preface into the book, stating that the author was not maintaining that the earth actually moved around the sun . . . just that this was a good hypothesis on which to base efficient mathematical models. Because the preface was unsigned, many assumed that it was written by Copernicus, therefore adding credibility to Osiander's contention that the idea of a heliocentric universe was only a suggestion, and thus serving to disguise the message.

History is divided as to the effect of Osiander's actions. Some write that this preface soft-pedaled the startling information, and as a result, scientists set about working directly with the new hypothesis without bothering to react because they viewed it as a hypothesis, not as a truth. Others write that in doing what he did, Osiander obliterated Copernicus's message. Whatever the effect, by 1543, when the book was published, Copernicus was dying. It is reported that he received a copy of the published book on his deathbed, but we will never know if he actually did.

From today's vantage point it is easy to shrug our shoulders and say, "What idiots! Why didn't they quickly realize Copernicus was right?"

But to the people of the sixteenth century, the thought of a revolving earth was nonsensical, and Copernicus's idea did not gain traction. These were primitive times scientifically. All observations of the sky were made by the naked eye—there were no telescopes and no advanced scientific equipment. If the earth was revolving, as Copernicus suggested, why didn't people fall off? Or if they overcame that concern, then they wondered why, if an arrow is fired vertically in the air, does it drop to the ground at the very place from which it was fired, as Aristotle had pointed out? To Copernicus's contemporaries, this proved that the ground had not moved while the arrow was in flight. The opposition was enormous and included intellectuals of the time, the day's religious leaders, and most important—adherents to Biblical teachings, for it was interpreted that the scriptures assert that the sun "turns" around an immobile earth.

Dead but Not Forgotten

But Copernicus and his ideas were destined not to be forgotten. Danish nobleman and astronomer Tycho Brahe (1546–1601) made strides in the field that permitted others to continue to explore these theories.

From adolescence, Tycho took an interest in astronomy. His observations of a nova (1572) and a comet (1577) began to teach him about the planetary system. At the time, astronomers believed that each planet revolved within its own sphere, but Tycho concluded that if a comet could pass through the heavens so effortlessly, then the spheres that were thought to carry the planets around the central earth likely did not exist at all.

Tycho's work became known to the Danish king, who gave him money to build an observatory and the island of Hveen, where he could build it. Tycho created greatly improved instruments and used a quadrant (an instrument for measuring the altitude of celestial bodies, using a 90-degree graduated arc with a movable radius) to record the position of everything from two angles. Tycho even took into account the variations of his materials when the metal shrank a bit during cold Danish nights. He eventually invented a sextant (a navigational instrument containing a graduated 60-degree arc, used for measuring the angular distances of celestial bodies to determine latitude and longitude), which provided even more accuracy.

Tycho also instituted the practice of observing something more than once and using different instruments to do so. (In contrast, Copernicus spent very little time observing the heavens and relied primarily on observations made by others.)

Tycho disagreed with Copernicus about the movement of the earth. In his opinion, his measurements showed that the earth was at rest with the moon and sun orbiting around it. The other five planets were satellites to the sun, which carried them along in their own orbit of the earth.

Tycho Brahe died in 1601, and his work was left to his assistant, Johannes Kepler (1571–1630), a skilled German mathematician who had joined Tycho two years before. Kepler inherited from the astronomer a wealth of the most accurate raw data ever collected on the positions of the planets, and in his work, Kepler came upon some new discoveries. It was Kepler who deduced that the planets move in elliptical orbits (the law of ellipses) rather than the circular ones that had been assumed by all who preceded him. Today we know that planets do not move in perfect ellipses because of the gravitational pull of other planets. However, Kepler's discovery was very significant because it broke the spell of circularity. Kepler's other two laws were the law of equal areas—an imaginary line drawn from the center of the sun to the center of the planet will sweep out equal areas in equal intervals of time—and the law of periods—the square of a planet's period (the time required for a planet to orbit the sun) is proportional to the cube of the semimajor axis of its orbit.

Kepler went on to write *Harmony of the World*, which explained the arithmetic of Copernicus's discovery that the farther a planet is from the sun, the longer it takes to complete an orbit. Then in 1618, 1620, and 1621, Kepler set out his *Epitome of Copernican Astronomy* theories.

In 1631, a year after Kepler's death, a French astronomer, Pierre Gassendi (1592–1655), became the first observer in history to see Mercury crossing the face of the sun, fulfilling a prediction made by Kepler. The result of this revelation was that Kepler's work and his theories were taken very seriously.

Bruno and the Copernican Message

While only a relative few knew of Copernicus and his message, his influence was not limited to scientists. Giordano Bruno was an Italian Renaissance philosopher who incorporated the theories of Copernicus into his writing. As a philosopher, and not a scientist, he ignored the mathematical computations that were the focus of the astronomers and wrote about his belief that the universe was infinite (a very different concept for that time) and that the stars were actually other suns spanning the infinite reaches of space. He also firmly believed in Copernicus's theory of a heliocentric universe. He advanced the Copernican view that the earth was not the center of the universe and believed that the earth revolves and that the diurnal rotation of the heavens is an illusion caused by the rotation of the earth around its axis.

Bruno taught and traveled throughout Europe, but he was a controversial figure who seemed to find trouble wherever he went. In 1591, after the death of the conservative Pope Sixtus V, Bruno felt the Inquisition was losing its strength and that it would be safe for him, who had been outspoken in his criticisms of the Church, to return to Italy. Unfortunately for Bruno, he was still a wanted man; he was arrested in 1592 and extradited to Rome, where he was imprisoned for six years. He was finally given an audience and was offered the opportunity to recant his beliefs, which included support of Copernicus. When Bruno refused to back down, he was declared a heretic, turned over to secular authorities, and burned at the stake on February 17, 1600.

While science has long claimed that Bruno died because of his Copernican beliefs, the actual charge against Bruno concerned his teachings that Jesus did not have a physical body and that his physical presence was an illusion. Though there is little doubt that his Copernicanism was a factor in his heresy trial, it was not the main issue.

In 1603 all of Bruno's writings were placed by the Church on a list of prohibited works; they remained on that list for almost four hundred years.

Galileo Proves His Point

While Bruno offended others in many ways, Galileo Galilei (1564–1642), who made great strides for the field of science, was marked for punishment for just one thing—and that was for proving that Copernicus was right.

Galileo is sometimes credited for inventing the telescope, but what he did was extend the use of it. Scientists in the Netherlands created the telescope, and Galileo refined it. Some instruments he created magnified up to 30 times. With these more powerful tools he clarified Aristotle's theory that the Milky Way consisted of stars and discovered that Jupiter had up to four satellites, which bolstered the theory that Earth was

Galileo.
Source: Library of Congress.

carrying a satellite (the moon) and that all objects did not revolve around Earth. However, his observations of Venus and its movement were the discoveries that lent support to Copernicus's theory that the planets orbited the sun.

Though seventy-five years had passed, people really didn't understand what Copernicus had said, so Galileo took it upon himself to spread the word of his own discoveries and relate them to those of Copernicus. He published his news in *The Starry Messenger.* Many were dubious about information gleaned by looking through a telescope, but Galileo promoted Copernicus's theories.

In *The Starry Messenger* and later in *Letters on Sunspots* (1613), Galileo was careful not to be too strident regarding his support of Copernicus. A devout Catholic, he had no interest in offending his patrons. In this post-Reformation era, it was

Source: *The Iconographic Encyclopaedia of Science, Literature, and Art* (1851).

easy to make enemies, and Galileo made no secret of his belief. They accused him of denying the truths of the scripture in declaring that the sun was at rest.

"Scandal!"

Events took a turn for the worse for Galileo. In 1616 the Inquisition warned Galileo not to hold or defend the hypothesis asserted in Copernicus's *On the Revolutions* because it was heretical, refuting a strict biblical interpretation of the Creation that "God fixed the Earth upon its foundation, not to be moved forever." Galileo agreed to keep mum, and things quieted down until 1623, when a close friend of Galileo's became pope and seemed to give vague permission to Galileo to ignore the ban and write a book on his beliefs. The result was *Dialogue on the Two Great World Systems*, which involved an argument between two intellectuals—one proposing a geocentric universe and the other a heliocentric one, and a layman who was neutral but interested. While the book presented the Church's geocentric view, the character who espoused it won Galileo no religious friends as he was portrayed as a fool.

The book was published in 1632 with the approval of Catholic censors, but the Church soon became angry, and Galileo was summoned to trial before the Roman Inquisition in 1633. Through a long and protracted series of interrogations, Galileo defended himself by saying that scientific research and the Christian faith were not mutually exclusive and that study of the natural world would promote understanding and interpretation of the scriptures. But his views were judged false. The Inquisitors eventually convinced Galileo to renounce all that he had written about his Copernican beliefs, and as a result, out of deference to his age, his ill health, and his agreeing to their terms, Galileo found himself under house arrest for his remaining years. His writings on the topic were added to the Church's list of prohibited works.

Which Brings Us to the Twentieth Century

And now amazingly, we must continue this story in the late twentieth century. In 1992, John Paul II established a commission to study the Inquisition's actions against Galileo. An article in the *New York Times* on October 31, 1992, states, "More than 350 years after the Roman Catholic Church condemned Galileo, Pope John Paul II is poised to rectify one of the Church's most infamous wrongs—the persecution of the Italian astronomer and physicist for proving the Earth moves around the Sun" (Alan Cowell, "After 350 Years, Vatican Says Galileo Was Right: It Moves," *New York Times*, October 31, 1992).

"We today know that Galileo was right in adopting the Copernican astronomical theory," the *Times* quotes Paul Cardinal Poupard, the head of the investigation, as saying.

John Paul II also attempted to rectify another wrong: Under his leadership, the Church put forward an official expression of "profound sorrow" and acknowledgment of error at Bruno's condemnation to death.

While it took at least a century for the Copernican revolution to be recognized, there is no doubt as to Copernicus's effect on the world of science. Though he did not do it single-handedly, ultimately Copernicus transformed humanity's thinking about its place in the universe. And that, indeed, is a revolution.

Copernicus's Skeletal Remains Just Unearthed

In 2005, archaeologists announced that what they believe are Copernicus's remains were found beneath the altar of a Polish cathedral. A computer reconstruction of the remains shows the head of a man of about seventy, Copernicus's age when he died in 1543, and a scar and a broken nose matched those visible in the astronomer's portraits.

11

How Isaac Newton Changed Our View of the Universe

Isaac Newton earned his place as the father of modern science by totally changing the way science was viewed. Newton's conceptions of gravity and mechanics, though now known to be not entirely correct, represented an enormous step forward in the evolution of human understanding of the universe. (Einstein moved understanding forward with his theory of relativity; for more about this, see chapter 13.) As a result, he is generally considered one of history's greatest scientists.

Newton not only developed revolutionary theories about motion, but he also conceived of a universal law of gravity and contributed to what we know about light and color. In his spare time, Newton created a new form of mathematics because he realized that he could not use the mathematics of his day to calculate such things as the movements of the planets. What he called the method of fluxions and we know as calculus is now used in every branch of the physical sciences, engineering, economics, business, and medicine. (If you read more about Newton, you'll read about a contemporary of his, Gottfried Leibniz, who lived from 1646 to 1716 and was also doing work on calculus. He denied that Newton invented it.)

How Newton Came to Study Science

Isaac Newton (1642–1727) was born into a family of farmers in Lincolnshire County, England. His father died three months before he was born, and when Newton was only two, his mother went to live with her new husband, leaving Isaac to be raised by his grandmother. He was reunited with his mother after the death of his stepfather when he was ten. Newton at-

Isaac Newton.
Source: Library of Congress.

tended school but was an unremarkable student (his school reports described him as "idle" and "inattentive"). No one would have predicted that he would one day be recognized as one of the greatest minds of all times.

When he was seventeen, his family removed him from school and attempted to have him learn the family farming business. Newton begged to return to school, and in 1661 he enrolled in Trinity College in Cambridge. He was older than most of his fellow students, and though his family had money, he entered as a sizar, a student who receives scholarship money in exchange for acting as a servant to other students.

The first two years of study for all students at Cambridge were devoted to learning the teachings of Aristotle, but by the third year, students were allowed more academic freedom. Newton was introduced to the philosophy of Descartes (he invented analytical geometry and did early work on the laws of

Aristotle and His Long Shadow

Aristotle, who lived in the fourth century BC (384–322), wrote so well and so intelligently for his time—and for future times—that for 1,800 years scientists based their own work on his teachings. Aristotle wrote on a vast number of topics, and among them were his views that the universe was geocentric (that everything revolved around the earth). As you read in chapter 10, during the seventeenth century, people were wrestling with this issue. Copernicus, Brahe, Kepler, and Galileo came to believe that Aristotle was wrong and that the universe was heliocentric (that everything revolved around the sun), but they had not yet devised a method for proving their theories. Newton's mathematics of fluxion (calculus) was able to do this.

In addition, Newton was able to provide new thinking on gravity that overrode Aristotle's conception of the elements. According to Aristotle, who knew nothing of gravity, the movement of nature was rooted in the fact that all things were made of the four elements: earth, air, fire, and water. Based on this, he predicted that "like would attract like," so items with more earth in them would find each other, and fire, on the other hand, would be repelled by earth, explaining why fire rises upward.

Newton's ability to move science forward—beyond and away from Aristotle—and to prove his theories were among his major contributions.

---

motion), Gassendi (he was among those who were revolting against Aristotle's teachings and was a very early believer in the existence of atoms or "matter in motion"), Hobbes (though best known for his political philosophy of civil government, Hobbes was significant to Newton for his rejection of Aristotelian and scholastic philosophy in favor of the "new" philosophy of Galileo and Gassendi), and Boyle (he founded the study of chemistry as a separate science and aimed to "improve natural knowledge by experiment"). Newton also studied the mechanics of Copernican astronomy, and Galileo's and Kepler's work as well. In one of Newton's notebooks, he wrote a statement in Latin that meant "Plato is my friend, Aristotle is my friend, but my best friend is truth."

History books report that science has the Great Plague, which struck England in the mid-1660s, to thank for Newton's best work. One out of every seven people living in London were dying from the bubonic plague, so in the summer of 1665, the university sent everyone home to try to halt the spread of the disease, and Newton returned to Lincolnshire. During this time of solitary study, while he was still less than twenty-five years old, he laid the groundwork for his revolutionary advances in mathematics, optics, physics, and astronomy.

Newton's Laws

In his work, Newton developed three laws of motion and proved that they apply to all that moves. To formulate these laws, Newton was looking for patterns, just as scientists before him had done. Newton, however, divided all physical motion into two separate categories: uniform motion, the movement of an object traveling in one direction at a constant speed or an object at rest, and acceleration, which applied to anything that changed direction or changed speeds. Planets fell into this second category.

These are his three laws:

1. **Every body continues in its state of rest or of uniform motion in a straight line unless it is compelled to change that state by forces impressed upon it.**
   Changes in motion do not happen spontaneously. There is always a reason for a change. The ball will continue to roll, and the book will continue to sit on the desk (in a continued state of inertia), until a force acts upon it. The ball stops when you put your foot in front of it or when friction causes it to stop, and the book falls when you push it.

2. **Force equals mass times acceleration.**
   This law defines the relationship between an object's mass, its acceleration, and the forces exerted on it ($F = ma$). This law can be used to determine the speed of anything from a cannonball to a thrown baseball to a spaceship and states that the greater the total or net force acting on an object, the greater the acceleration.

3. **For every action there is an equal and opposite reaction.**
   Basically this law states that all motion comes in equal and opposite pairs. When you swim, you push the water

backward. The water reacts by pushing you forward. While it is the best-known law, it is also the least intuitive since we never think about the water pushing against us as we paddle. In 1904, the unit of force was named the newton in his honor. A newton is defined as a unit of force equal to the force that imparts an acceleration of 1 minute per second per second to a mass of 1 kilogram; it is equal to 100,000 dynes (measurement that uses centimeters, grams, and seconds) or a bit less than a quarter of a pound, which equals the weight of a medium-sized apple.

Because Newton not only created the laws but created the mathematical language to make calculations using these laws, he created the ability to quantify the natural world. As a result, this scientific method can be used to derive exact equations that solve specific problems—from how to build a skyscraper to how to create a theme park ride.

Newton's Law of Universal Gravitation

Newton also redefined the concept of gravity. While German astronomer Johannes Kepler (1571–1630) had empirically devised the mathematical foundation for describing planetary motions (his laws of planetary motion), and Galileo Galilei had presented empirical relationships for how things fall (early studies in gravity and motion), these areas—celestial and terrestrial—were seen as two separate fields.

Newton was the one who realized that Galileo and Kepler were studying the same thing, and he developed his law of universal gravitation: Between any two objects there is an attractive force proportional to the product of the two masses divided by the square of the distance between them. In other words, an apple falling to the ground and the moon orbiting the earth are both affected by gravity in the same way.

The very idea that an arrow, a thrown ball, the planets, and our own blood were all controlled by the same laws of motion was a truly revolutionary thought. Newton's work meant an end to much of the trial-and-error approach to learning things, because suddenly nature was predictable in such a way that experiments could now prove or disprove hypotheses.

With the law of universal gravitation, Newton closed the circle on his work. He had the force—gravity—that operated everywhere, and he had the rules—the laws of motion—that governed the operation of all forces. Suddenly with these theories and the matching formulas, scientists could begin to predict everything in a whole new way.

One of the first to successfully employ Newton's laws was Edmond Halley (1656–1742) of Halley's Comet fame. Using historical records and Newton's laws, Halley worked out the orbit of the comet that now bears his name. When the comet actually reappeared on Christmas Day in 1758, as Halley predicted it would (unfortunately, Halley died before this happened), the event powerfully underscored the idea of the clockwork universe and proved that reliable predictions were possible.

Today we know that the clockwork universe isn't totally predictable. Although Newton's mechanics correctly predicts how planets and everyday objects behave, it did not correctly explain the behavior of very small objects such as atoms, nor did it explain the behavior of things traveling near the speed of light. In the 1920s a new type of mechanics (quantum mechanics) evolved to further explain the universe.

Chaos theory is another new challenge to Newton's theories of predictability. With certain systems, the initial conditions can rarely be measured accurately enough to predict their behavior at future times. For example, with weather, no matter how much meteorologists measure wind speed, air temperature, and barometric pressure, they are never able to predict exactly what time it will start raining tomorrow, let alone what the weather will be a year from now. The chaotic nature of atmospheric motion makes it difficult to predict.

However, Newton's development of the clockwork universe was the first example of the scientific method in use in which there was an interplay of observation and theory that led to new theories and experiments that modified existing theories.

Writing *Principia*

Newton's master work, *Philosophiae Naturalis Principia Mathematica* or *Mathematical Principles of Natural Philosophy*, better known as *Principia*, might never have been written—and certainly wouldn't have been published—if it hadn't been for Edmond Halley. In 1684 Halley paid a visit to Newton in Cambridge and posed a mathematical question to Newton about the planets. Newton could answer the question but couldn't locate the piece of paper on which he had written his notations about the calculation. Halley encouraged him to write a paper on the topic, and Newton did much more. He retired for two years and produced his masterpiece, *Principia*. When the Royal Society of London, which had intended to publish it, backed out because of financial difficulties, Halley had to reach into his own pocket to pay for the publishing (1687). The result was instant fame for Newton—and a major step forward for science.

Principia put forth Newton's three laws of motion (a body will not change its state of motion unless acted upon by another force; a body will keep moving in a straight line until some other force acts to change its speed or deflect it; and every action has an opposite and equal reaction). He also put forth his universal law of gravitation (every object in the universe exerts a tug on every other).

For a genius who had exerted such an amazing influence on science, Newton made most of his scientific contributions early in life. Later, he devoted time to other things, including alchemy (a medieval "science" that involved trying to change base metals into gold). Then in the 1690s he wrote a number of religious tracts and served as a member of Parliament in 1689–1690. He moved to London in 1696 and became warden and eventually master of the Royal Mint, where he oversaw recoining and also went after coin clippers (an illegal activity) and counterfeiters. Ironically, it was his work at the Mint, rather than his contributions to science, that earned him knighthood (1705).

Newton's laws of motion and gravity provided a basis for predicting a wide variety of different scientific or engineering situations, especially the motion of celestial bodies. His calculus proved vitally important to the development of further scientific theories. Finally, he unified into a satisfying system of laws many of the isolated physics facts that were being discovered.

Albert Einstein's theory of general relativity, the current best theory of gravitation, incorporates Newton, Kepler, and Galileo and takes their work forward. In the future, some physicist will produce a final unified field theory that takes everything another step further. True revolutions are rare in science. Things evolve.

12

Atomic Theory

How They Discovered Something Way Too Small to Be Seen

Though none of us wake in the morning and think about how our pajamas, our beds, our floors, our houses, and our breakfasts are all made up of atoms, it's something that we all just "know" and know that someplace along the way, we learned it in science class. When Dupont used to run television commercials that promised "better living through chemistry," we all had an innate understanding that chemistry was possible because scientists could break things down into atoms and reassemble them differently.

In many ways the science of atoms is still in its infancy. It was only one hundred years ago—during what is known as Einstein's "miracle year" (see chapter 13)—that scientists verified the fact that atoms are real. But despite the relative youth of the science, our everyday lives are changed immensely by what they have learned so far. Your daily encounters with the miracles of chemistry would fill many pages of this book, so suffice it to say that the multivitamin you take in the morning, the touch of spandex in the slacks you put on for the day, the chemical mix of the gasoline that powers your car, the pressed-wood compound your desk is made of, and the delicious cake you or a family member mixes up for dessert on the weekend are all available because of an understanding of chemistry. If we shift focus away from our own lives, we must think of the chemical composition of the space shuttle skin, the miracle creation of innumerable medicines, and the beauty of the fireworks displays on the Fourth of July.

Since humans began using fire 1.5 million years ago, we have known how to produce and control chemical reactions. What we lacked was an understanding of how they worked. It took the discovery of atoms—nature's building blocks—to begin to develop a blueprint for how those chemical reactions can be made to work even more effectively. Atomic theory, the discovery and understanding of atoms and how they interact, is one of the most important theories in the history of science. Yet how do you conceive of or discover something that is way too small to be seen?

Those Greeks Again

From early on, people obviously understood that substances you could eat were different from those you might use to build a shelter, but they lacked an understanding of what those substances themselves could consist of. A fifth-century BC Greek philosopher and scientist, Empedocles, felt that all matter was composed of four elements (or as he called them, "roots"): fire, air, water, and earth, and it was the ratio of the mix of these elements that caused differentiation. A rabbit had more water and fire in it, causing it to be soft and have life, while a hard and inanimate stone must be made up primarily of the element of earth. The acknowledgment that substances—even ones like stone that appeared "pure"—were made of a combination of elements was a big step forward in the field of science.

Just a few decades later, another Greek came up with an idea that could have transformed the world, but the world wasn't ready for it. Democritus (460–370 BC) realized that one of the problems with Empedocles' theory was that no matter how many times you split a rock, you never came up with anything that resembled any of the elements. Democritus suggested that if you were to continue cutting a stone into smaller and smaller pieces, you would eventually cut to a point where the tiny pieces were so small that they could no longer be divided. Democritus called these tiny pieces *atomos*, meaning "indivisible," and theorized that the *atomos* of stone were unique to stone, and the *atomos* of fur were unique to fur.

Unfortunately for the world, Aristotle and Plato did not agree with Democritus. Influential Aristotle was troubled by the lack of proof, and he pushed forward Empedocles' idea

What You May Have Forgotten from Science Class

Today we have a very different scientific definition for the word *element*. As early as 1660, Robert Boyle (1627–1691) recognized that the Greek definition of element (earth, fire, air, and water) was not correct. Boyle proposed a new definition of an element as a fundamental substance, and today we define elements as fundamental substances consisting of atoms that all have the same number of protons and that cannot be broken down further by chemical means.

An *atom* today is defined as a single unit of an element. The atom is the most basic unit of the matter that makes up everything in the world around us, and each atom retains all of the chemical and physical properties of its parent element. We now know that atoms consist of positively charged *protons*, neutral *neutrons*, and negatively charged *electrons*.

Most things around us are *compounds*, substances formed by the chemical combination of two or more kinds of atoms. The smallest particle of a compound is a *molecule*, which consists of two or more atoms.

about the elements—fire, air, water, earth—as more likely. As a result, Democritus's theory of the *atomos* never gained traction.

Getting Closer

Two thousand years had to pass before anyone revisited in a helpful way the question about what things were made of. It was actually a pupil of Galileo who made one of the relevant discoveries.

In 1643 Evangelista Torricelli, an Italian mathematician, discovered that air had weight—he showed that air was capable of pushing down on a column of liquid mercury (thus inventing the barometer). Then a Swiss mathematician, Daniel Bernoulli (1700–1782), undertook the study of the way air moved over a bird's wing. He determined that air and other gases could provide lift to the wing because they consist of tiny particles that are too small to be seen. He surmised that the particles were loosely packed in an empty volume of space and that they could not be felt because the tiny particles moved aside when a human hand or body moved through them. He reasoned that if these particles were not in constant motion, they would settle like dust, so he suggested that they were continuously moving and bouncing off each other. (Smart thinking!)

In 1773 Englishman Joseph Priestley found that when heated, mercury calx, a red solid stone, could be turned into two substances—a silver liquid metal and a gas. This was key evidence necessary for scientists to begin to understand that substances could combine together or break apart to form new substances with different properties. The question was when the substances were broken down, what were the basic building blocks? No one had yet determined that.

The Fog Lifts on Atomic Theory

That was left to John Dalton (1766–1844), a British teacher who specialized in observing weather, who put forward the first modern atomic theory. Dalton's observation of the weather actually led him to develop his theory. Dalton saw that water, in the form of fog, could exist as a gas that mixed with air, yet when water was frozen into ice, it was a solid and had to exist separately. Dalton wondered, why could water sometimes behave as a solid and sometimes as a gas? He performed a series of experiments on mixtures of gases to determine what effect the properties of the individual gases had on the properties of the mixture as a whole, and he came to believe in the existence of atoms because of these experiments.

He also began to see that most things could be broken down by burning or immersion in acid or some other procedure. When he ran into something that could not be broken down, Dalton called these *elements* (oxygen, gold, sulfur, iron, etc.). Dalton soon began to realize that chemicals have a precise ratio of elements. Water—no matter where it comes from—always has ratio of 2:1 hydrogen to oxygen. Dalton guessed that each chemical element is represented by its own atom and that these atoms combine in simple ways. Water (H_2O) is made from two atoms of hydrogen and one of oxygen, for example.

Dalton came to the conclusion that Bernoulli's theory of tiny particles was correct, but that the theory applied to all matter—gases, solids, and liquids. Though he first floated his ideas out in 1803, his paper on the topic, *A New System of Chemical Philosophy*, was not published until 1808. By that time, Dalton had developed four main concepts of atomic theory:

1. All matter is composed of indivisible particles called atoms.
2. All atoms of a given element are identical—and have the same mass; atoms of different elements have different properties.
3. Chemical reactions involve the rearrangement of combinations of those atoms, not the destruction of atoms.
4. When elements react to form compounds, they react in defined, whole-number ratios.

While some of the details of Dalton's original atomic theory are now known to be incorrect, the core concepts of the theory—that chemical reactions can be explained by the union and separation of atoms, and that these atoms have characteristic properties—are the foundations of modern science. Like so many new ideas in science, it took about fifty years before everyone accepted this theory.

More Discoveries Are Made

Even as scientists were agreeing on atomic theory, they were realizing that atoms were made of still smaller substances.

Until the end of the nineteenth century, the accepted model of the atom was that of a billiard ball, a small solid sphere. In 1897 J. J. Thompson (1856–1940) discovered the electron. Working with cathode ray tubes, Thompson discovered that when an electric current passed through the tube, a stream of glowing material was visible. He noticed that if a positively charged electric plate was nearby, the glowing stream would bend toward it. Thompson decided the stream was actually made up of small particles, pieces of atoms that carried a negative charge (electrons). This discovery dramatically changed the modern view of the atom and became known as the "plum pudding model," which Thompson proposed before the discovery of the proton or the neutron. In this model, the atom is composed of electrons surrounded by a soup of positive charge, like plums surrounded by pudding. The electrons were thought to be positioned uniformly throughout the atom. This discovery dramatically changed the modern view of the atom. While it pushed science a step forward, it was soon to be disproved by Rutherford's gold foil experiment (see below).

They Actually Are Real!

Despite all that scientists were learning, there was still no way to see atoms, so the debate continued as to whether atoms were theory or real. Finally Albert Einstein (see chapter 13) ended this debate in 1905 when he explained a phenomenon called Brownian motion. When a small particle such as a grain of pollen is suspended in a liquid and observed under a microscope, it is seen to move around in a random path. Einstein explained that the particle moved because of collisions with atoms. Einstein was able to explain it mathematically.

It wasn't until 1980, however, that any one could document this visually. "Smile!" must have been the operative word in the laboratory at the University of Heidelberg in Germany, where that first photograph of an individual atom was finally able to be produced.

The Significance of the Gold Foil Experiment

In 1908 Ernest Rutherford performed a series of experiments with radioactive alpha particles in what became known as his "gold foil experiment." When he fired tiny alpha particles at solid objects such as gold foil, he found that while most of the alpha particles passed right through the gold foil, a small number of alpha particles passed through at an angle, as if they had been pushed off a straight line path, and a smaller number bounced straight back, as if they had bounced against a wall. Rutherford's experiment suggested that gold foil—and matter in general—has holes in it since it lets most things pass through while a small number ricochet off.

Then in 1911 Rutherford suggested that atoms consist of a

Ernest Rutherford.
Source: Library of Congress.

small dense core of positively charged particles in the center or nucleus of the atom, surrounded by a swirling ring of electrons. The nucleus is so dense and positively charged that positively charged alpha particles bounce off it, but the electrons are so tiny and spread out at such great distances from the nucleus that the alpha particle passes through that part of the atom. Rutherford's view resembled a tiny solar system with the positively charged nucleus always at the center and the electrons revolving around the nucleus. (The positively charged particles in the nucleus were called protons, and they carried a charge equal to but opposite that of the electrons.) To provide a concept of the amount of empty space within an atom, teachers sometimes use the Yankee Stadium analogy: If a ball (smaller than the size of a baseball) is placed on the pitcher's mound and represents the nucleus of a hydrogen atom, then you would have to walk all the way to the edge of the stadium to reach the electron.

Then in 1932 James Chadwick (1891–1974) discovered a third type of subatomic particle, which he named the neutron. Neutrons help stabilize the protons in the atom's nucleus. Because the nucleus is so tightly packed together, the positively charge protons normally would repel each other. Neutrons help to reduce this repulsion and stabilize the atom's nucleus. They are always electrically neutral.

In 1915 Niels Bohr (1885–1962) further advanced atomic theory by developing the Bohr model of the atom, which soon led to the development of the modern quantum theory.

Next scientists began to suspect that atoms might be made from even smaller particles called *quarks*, and the debate over that issue mirrored the old argument about atoms. But that's another story.

Never Say Scientists Lack a Sense of Humor!

A neutron walked into a bar and asked the price of a drink. "For you," the bartender replied, "no charge."

What did the proton say to the neutron?
"Try to be more positive."

What did the neutron say to the electron?
"You're always so negative."

The Characteristics of an Atom

The weight of an atom is roughly determined by the total number of protons and neutrons in the atom. While protons and neutrons are about the same size, the electron is more than 1,800 times smaller than these two.

Atoms of different elements are distinguished from each other by the number of protons they have (the number of pro-

A Dangerous Profession

Learning about chemicals was a dangerous process. George and Thomas Knox suffered when they were exposed to hydrofluoric acid. Thomas almost died, and George spent three years recovering his health.

George Gore of London isolated a small quantity of fluorine, but it exploded and destroyed his laboratory. Jerome Nickels of France and Pauline Louyet of Belgium died in their laboratory, asphyxiated by the gas. Lungs and skin were blistered.

There is little doubt that Marie Curie's (1867–1934) work with radioactivity greatly affected her health. Marie died from leukemia at sixty-seven, her fingers burnt by "her dear radium."

We owe a great debt to those who risked their lives to learn about chemical elements.

tons is constant for all the atoms of a single element; the number of neutrons and electrons can vary under some circumstances).

Organizing the Elements: The Beauty of the Periodic Table

By the mid-nineteenth century, 63 elements had been identified. While gold, copper, silver, tin, lead, and mercury had been known since antiquity, in 1669 an alchemist's discovery of phosphorous, the first element to be discovered in more modern times, put the discovery process on a roll, and an increasing number of elements were added to the list after that. Then in the 1780s, Antoine Lavoisier (1743–1794) ushered in a new era of chemistry by making careful quantitative measures that allowed the compositions of compounds to be determined with accuracy. He also divided the few elements known in the 1700s into four classes.

Scientists were beginning to realize they needed an organizational system for keeping track of the elements. Carl Linnaeus (1707–1778) had organized species into categories, so scientists were determined to bring some type of order to the chemical elements.

While this seemed like a good idea, no one was quite certain how to do it. What did an element like oxygen, a gas, have in common with an element like mercury, a liquid, or a soft metal like platinum, or the ones that were too dangerous to handle without gloves—fluorine and potassium?

John Dalton was working on a simple table, and fewer elements were known in the early 1800s. The very first periodic table was probably created by a French geologist, A. E. Beguyer de Chancourtois, in 1862. While there were major flaws in his attempt, he recognized that elemental properties recur with every 7 elements—what is meant by periodicity.

Dimitri Mendeleev (1834–1907) is generally regarded as the primary creator of the periodic table. He wrestled with how to organize the 63 elements, and when he finally figured it out, he arranged the elements into an orderly table, which he published in 1869. Today the vertical columns are called *groups* and the horizontal rows are called *periods.* His organizational system involved arranging the elements according to their atomic weights, and he noted the following:

- Elements with similar chemical properties have atomic weights of nearly the same value.
- The arrangement of the elements in the order of their atomic weights is based on their distinctive chemical properties.
- The elements that are the most widely diffused (have more space within them) have small atomic weights.
- The magnitude of the atomic weight determines the character of the element.

Mendeleev also forecast the properties of elements yet to be discovered based on the properties the periodic table dictated that they would have.

At about this time, a German chemist, Lothar Meyer (1830–1895), had also prepared a table that in many respects resembled the present periodic table. He did not publish this work until after the appearance of Mendeleev's first paper on the subject in 1869. His table was very similar to that of

Mendeleev, but it contained some improvements and was, perhaps, influential in causing some of the revisions Mendeleev made in the second version of his table, published in 1870. In general, Meyer was more impressed by the periodicity of the physical properties of the elements, while Mendeleev saw more clearly the chemical consequences of the periodic law.

Q: Why are chemists great for solving problems?
A: Because they have all the solutions.

Old chemists never die, they just fail to react.

13

Einstein Made Clear (Enough)

If you are the type of person whose eyes glaze over at the mention of Albert Einstein or $E = mc^2$, then this is the Einstein chapter for you. You are treated to two interesting stories—one right away, and one you have to read the chapter to find. In the process, you're introduced to enough about Einstein that you'll have some understanding why he is regarded as one of the finest scientific minds the world has ever known.

"The important thing is not to stop questioning."
—Albert Einstein

And just in case you need a little more enticement, you should know that while we may not spend our days pondering Einstein's theory of relativity or Brownian motion, Einstein is actually with us every day because of the vast number of technological inventions he made possible—everything from GPS navigators to laser technology. His work laid the foundation for the invention of computers, clear television pictures, CDs, DVDs, and so much more. Our digital cameras are also due to Einstein. They contain small sensors that convert light into electricity, and this is possible because of Einstein and his work with the photoelectric effect. (See below.) He deserves a quick read!

Albert Einstein (1879–1955)

The first story about Einstein is for every student who has ever struggled or been miserable at school, and for all parents who ever fretted about whether their children would ever make anything of themselves.

The child who was to grow up to be overwhelmingly acknowledged as the greatest scientist of the twentieth century was a late talker and was considered a slow learner. It is said that until he was seven, he said each sentence quietly to himself before saying it out loud. While he pored over science and math books at home and his grades were good, his teachers found him to be an unremarkable student.

When Einstein was a teenager, his parents were forced to move to Milan to find work, and they placed him in a boarding school to complete *gymnasium* (what we consider high school). Without telling his parents, he quit a year and a half early—only to get his comeuppance a bit later when he was required to take a qualifying exam to enter the Swiss Federal Institute of Technology in Zurich. Einstein excelled on the math and science portion but failed the liberal arts section. One can imagine the family frustration, stress, and concern when his parents then had to send him to Aarau, Switzerland, to finally finish secondary school (1896). At about this time—at age sixteen—Einstein renounced his German citizenship because he refused to participate in the government requirement of mandatory military service.

Even after graduation, Einstein's life was proceeding by fits and starts, and he was initially unable to find work. Two fellow graduates were able to line up teaching jobs, but no one was interested in hiring Einstein. Finally the father of a classmate helped him get a job as a technical assistant examiner at the Swiss Patent Office. Though he was able to maintain employment, in 1903 he was initially passed over for a promotion until he had "fully mastered machine technology."

In later years, Einstein did not focus on his past setbacks. Reports have it that he remembered two significant facts about his upbringing. The first is a lesson in good parent-

Albert Einstein.
Source: Library of Congress.

ing—Einstein's parents introduced him to things that captured his interest: A compass that Einstein's father, a former featherbed salesman who later ran an electrochemical plant, showed five-year-old Einstein engaged the boy completely. He was fascinated by the thought that something in "empty" space caused the compass to react. Einstein's mother gave him another gift during these years—she encouraged his playing of the violin, a hobby that Einstein pursued throughout his lifetime.

And in what might be construed as a don't-rush-a-child message, Einstein is said to have attributed his development of the theory of relativity to his slowness, saying that by pondering space and time later than most children he was able to apply a more developed intellect to the issue.

What Einstein Did for Science

This would be a much shorter chapter if Einstein had made just one contribution to science. But despite his slow start in academics, Einstein went on to apply himself to many different scientific puzzles, and as a result, his contributions were many.

Just after submitting his doctoral thesis in 1905, Einstein went on to submit four significant articles that laid the foundation for modern physics. (The year became known as Einstein's *annus mirabilis*, or miracle year, and its centennial was widely celebrated in 2005.)

At the age of only twenty-six, Einstein wrote three spectacular scientific papers on three very different topics between the months of March and June of 1905. Each of the papers put forth a truly revolutionary theory that upended some area of science. Then in October of that same year, he added an astonishing postscript to the third paper, which is now the best-known work of all, for this was when he introduced his now-famous equation, $E = mc^2$.

No single person has contributed so much to science and done it so quickly since the late seventeenth century, when Newton discovered gravity, founded the science of optics, and invented calculus. And with this work, Einstein ushered in the modern era of science. Here's what this twenty-six-year-old patent clerk put forward.

Photoelectric Effect

Einstein's first paper for the year was on radiation and the energetic properties of light. A few years previously, German physicist Max Planck suggested that energy in an atom occurs in little chunks called *quanta*. Building on this in his paper, Einstein suggested that light also existed in chunks or quanta (now called *photons*). Einstein also found that he could release electrons from metals when high-frequency (ultraviolet) light was shined on them. This showed that light was acting like it was made of little particles and not like a wave. If it were a "light wave," the metal would just heat up due to the transfer of energy. Einstein's theory grew out of the realization that the emission of light by an atom is not a smooth continuous flow like the emission of sound from an instrument, such as a violin. Instead the release of light occurs more as a burst of energy.

Eventually this photon hypothesis turned out to be the key to unlocking the structure of the atom, and the paper became the foundation of quantum physics (the language of the atom). In 1921 he earned the Nobel Prize for Physics for it.

Between 1916 and 1925 Einstein made other contributions to the study of light, including the idea of the stimulated emission of radiation—a concept that led to the development of the laser.

Atoms Exist! (Brownian Motion)

Throughout the nineteenth and early twentieth centuries, scientists debated whether atoms were theory or real, and since atoms were too small to see, the arguments went unresolved. Einstein ended this debate in May of 1905 when he submitted a paper explaining Brownian motion. This phenomenon was named after English physician Robert Brown, who in 1828 noticed that microscopic grains of pollen floating on the surface of water continued to move randomly, long after the water had become stationary. Einstein had puzzled over this while working on his doctoral dissertation, and he began to imagine that the pollen was like a bowling ball being pummeled from all direction by water molecules. Einstein explained that the particles moved because of collisions between atoms—and he put forward that this proved that atoms were real because they were causing "real" things to happen.

Three years later, French physicist Jean Perin succeeded in creating a measuring method that proved that Einstein was right, forever changing all of physics and chemistry as well as biology, because this also led to the science of molecular genetics.

Einstein was also proven right (but not until 1980) by the use of scanning tunneling microscopes, where photographs of individual atoms can be taken. (After the discovery of the atom, scientists began arguing about whether tiny particles inside atoms are really made from even smaller particles called *quarks*—or just act like they are. The debate mirrored the old argument about atoms, though today, quarks are deemed real.)

Special Theory of Relativity

In June, shortly after his Brownian motion paper, Einstein wrote his first paper on relativity. (A few months before Einstein completed his paper, the well-regarded French mathematician, Henry Poincaré, published on the topic.) *On the Electrodynamics of Moving Bodies* was an amazing document partly because of the ideas it introduced but also because it differed greatly from the way most scientific papers were presented: Einstein provided little reference to what had preceded it or what or who had influenced him to develop the theory; and in this particular paper, he had no footnotes and showed very little math.

The special theory of relativity changed scientists' conception of space and time. Einstein demonstrated that time was relative to the speed at which the observer is traveling. For example, imagine you are riding in a car, and you are observing another car. If both cars are traveling the same speed, say 30 miles per hour, then your perception of the other car is that it isn't moving, because it stays directly beside you as you progress—therefore, its speed relative to you is zero. Einstein went on to show that if from your perspective someone is moving, you see time elapsing more slowly for him than for you. (This is hard to grasp because at everyday speeds the slowing is less than one part in a trillion and is thus imperceptibly small, but it makes a big difference for scientists studying the universe.) Einstein also theorized that the speed light travels (which we understand to be 186,000 miles or 299,330 kilometers per second) is not absolute.

The special theory of relativity also explained that when matter is converted into energy, energy can be released in a predictable pattern. This leads directly to Einstein's well-known formula that proves that mass and energy are interchangeable.

When the paper presenting the special theory of relativity was first published, the equation was not presented along with it. It came along, almost as a footnote, in a separate short paper that was released in October.

$E = mc^2$

Einstein's final contribution for 1905 was *Does the Inertia of a Body Depend upon Its Energy Content?* This paper introduced the famous equation that the energy of a body at rest (E) equals its mass (m) times the speed of light (c) squared. (This equation was actually first published by Poincaré in 1900.) The formula implies that a small mass can be converted into a huge amount of energy, and vice versa.

In an op ed piece in the *New York Times*, on September 30, 2005, Brian Greene, a professor of physics and mathematics at Columbia University and a well-regarded author of many science books, brought up a little-known fact about the September 1905 paper. Einstein didn't actually write $E = mc^2$; he wrote the mathematical equivalent $m = E/c^2$, placing greater emphasis on creating mass from energy than on creating energy from mass.

Whether matter is coal (think of heat), gasoline (think of a moving car), or cinnamon bread (think calories), energy is within it. The most famous mathematical equation in history provided a formula that showed that at the atomic level, matter and energy can be converted into one another, and the equation explained the consequence of the theory of relativity. (Since speed of light squared is such a huge number, it shows that it is truly a lot of energy.)

In 1905 Einstein had no idea how that energy could be released or used, and though it took until the 1930s, Einstein's work—and this equation—permitted the discovery of nuclear fission—a way to release the energy stored in the nuclei of atoms by splitting them into smaller atoms. Eventually this was to provide a methodology for obtaining power from nuclear reactions using nuclear reactors and atomic bombs.

Expanding on Special Relativity

In 1910 Einstein began teaching, and from 1914–1919, he was director of the Kaiser Wilhelm Institute for Physics in Berlin. Despite the incredible changes he had brought to the world of science in 1905, Einstein still had more work to do. Since special relativity had to do with objects moving in straight lines at constant speeds, Einstein began to generalize his thinking to include curved paths and accelerating objects.

Einstein came up with the concept of general relativity in 1907, but it took eight years before he had worked out the math that explained it. When that happened, he told a friend "I was beyond myself with excitement."

In 1915 he gave a lecture in which he worked with an equation (first published by German mathematician David Hilbert) that replaced Newton's law of gravity. In general relativity, gravity is no longer a force (as it is in Newton's law of gravity) but is a consequence of the curvature of space-time and has the capacity to bend light. The general theory of relativity predicted that a light beam passing near a massive object (such as a planet) would actually be bent, and by how much. This prediction was supported during a total eclipse of the sun in May of 1919 and again by observations of a total eclipse in Australia in 1922.

General relativity predicts that the relative clicking of clocks changes depending on their position in a gravitational field. Satellites located high above the earth are moving in a gravitational field that is slightly weaker than what we experience on the ground, and as a result, their internal clocks tick at a different rate than those on earth. The effect is extremely small, but it is this level of accuracy that permits global positioning systems to work. (The time differences between the signals in your car and those from the satellite tracking you is accurate to about a billionth of a second.) Of course, in 1915, Einstein was not doing this work so that we could have global positioning devices in our cars. His work laid the foundation for the study of cosmology, led to the development of the idea that our universe was created in the big bang, and put forward what scientists are now proving—that our universe is still expanding. It gave scientists the tools needed for understanding the many features of the universe, including a greater understanding of quasars and black holes.

Because Einstein's theory was a combination of mathematical reasoning and rational analysis, as opposed to the usual scientific methodology of experiment and observation, many people disbelieved his idea. Then in 1919, a solar eclipse occurred that permitted another scientist, Arthur Eddington, to conduct measurements that tested Einstein's theory. He measured how much the light from a star was bent by the sun's gravity when it passed close to the sun (called *gravitational lensing*), and as a result, Einstein began to be recognized for what he had accomplished. However, when the Nobel Prize was given to him in 1921, it was for his photoelectric work—more widely accepted than his work on relativity. The Nobel committee felt the award should be given for the work that was clearly accepted by the scientific world of the day.

It is interesting to note that even in later years, Einstein did not work in a laboratory, he lacked staff, and he never thought of looking for research grants. He was a pencil-and-paper conceptual thinker who figured everything out in his head. Because his concepts were so advanced, his theories usually predated science's ability to verify his thinking; as a result, some of his ideas took a long time to be accepted, even among scientists.

Einstein Refrigerator

Most scientists are tinkerers, and Einstein, too, occasionally worked on the practical, not just the theoretical. In 1930, he and former student Leo Szilard received a patent for a refrigeration unit that featured thermodynamic refrigeration that cooled but required no moving parts.

Later Life

When the Nazis came to power in 1933, Einstein knew his time in Germany had ended. The Nazis made a concerted effort to discredit his theories and accused him of creating "Jewish physics." They also blacklisted anyone who taught Einstein's theories. As a result, Einstein renounced his German citizenship and fled to the United States. Many citizen groups in the United States accused Einstein of communism because of his belief in a democratic socialist system (a combination of a planned economy that had an underlying respect for human rights), so the government had quite a file on him. But because there were no negative government-generated documents, the United States offered him permanent residency.

In 1939, Einstein contacted President Franklin Delano Roosevelt about the possibility of American exploration of the use of nuclear fission for military purposes. Einstein feared that the Nazis were at work on this possibility and would harness nuclear fission first. Based on this—as well as advice from others—FDR launched the Manhattan Project, which of course, eventually built the atomic bomb. Though he suggested the investigation out of fear for what the Germans had or could do, Einstein was very much against the use of nuclear weapons and actively fought against nuclear tests and bombs.

Shortly after arriving in the United States, Einstein accepted a position at the newly founded Institute for Advanced Study in Princeton, New Jersey. At the Princeton Institute, Einstein worked on unifying the laws of physics—he referred to his work as the unified field theory. His goal was to simplify (by unifying) all the fundamental forces of the universe, but unfortunately, he was unsuccessful. Today scientists are still at work on this, but we may yet see the day when they discover a unified theory.

Einstein as a "Regular Guy"

The Einstein of later years was well remembered as a kind and decent pacifist. He also had a sense of humor. We've all seen the famous picture of him sticking out his tongue for the camera. It was taken in 1951 on his seventy-second birthday, when photographers had been hounding him all day. After being coaxed to smile one more time for the camera, he stuck out his tongue instead.

He enjoyed sailing and playing the violin—something his parents made him learn when he was a boy. He was very definitely the "absent-minded professor" and would become so rapt in solving physics problems that he would become oblivious to his surroundings.

And now, finally, you've earned the final story: Einstein spent his final years at Princeton apparently wearing the same outfit every day. While curious onlookers may have suspected he slept in his clothes, he actually had a closet full of multiple copies of the same suit to avoid daily sartorial dilemmas.

The elderly Einstein would often wander the Princeton campus muttering to himself. Lore has it that his coworkers, fearing he'd say something brilliant that would then be forgotten, hired a student to follow him at all times and take notes.

And here is a tantalizing thought. The nurse who was with him before he died said that just before he breathed his last breath, he mumbled several words in German.

Einstein's Brain

Despite the fact that few laypeople could describe his work, Albert Einstein was the rock star of scientists, and people were fascinated by him. Charlie Chaplin once said to him, "I am adored because . . . You are adored because you are incomprehensible."

After his death, his executors worried that if he were buried, curiosity seekers would forever create problems at the cemetery, so it was arranged for his body to be cremated, and the ashes were scattered in an undisclosed location. This all would have been well and good, except a pathologist carried off his brain to study it.

Dr. Thomas S. Harvey, a pathologist at Princeton Hospital, performed the autopsy and removed Einstein's brain at that time. (Einstein had said he did not object to the study of his brain, but he did not want the findings publicized.) Harvey was very protective of his unique "possession," and despite numerous career moves always carried the brain samples with him. (He did, however, fulfill some requests from researchers who approached him.)

Years later (in the 1970s), an editor for *The New Jersey Monthly* read something about the existence of Einstein's brain, and after making some inquiries, sent reporter Steven Levy out on the story. Levy tracked down Dr. Harvey, who was then living in Wichita, Kansas. Levy notes that though Harvey was reticent when the interview began, before Levy left, Harvey had gone to the office closet, pulled out a cardboard box marked "Costa Cider," and revealed two mason jars with sections of Einstein's brain within.

In 1996 Harvey finally brought the remaining pieces back to Princeton Hospital.

So is Einstein's brain any different from yours or mine? Recently Canadian researchers were given access to the recovered brain, and they found that he had an unusually large inferior parietal lobe—a center of mathematical thought and spatial imagery—and shorter connections between the frontal and temporal lobes.

"Only one who devotes himself to a cause with his whole strength and soul can be a true master."
—Albert Einstein

14

A Ticket to Pluto and Beyond

If you had been born just about one hundred years ago, one of the great discoveries of our solar system—that of locating the ninth planet—was still waiting to be made. And Clyde Tombaugh, the fellow who was to become the first American to discover a planet, was just a farm boy from Kansas who had only a high school diploma.

Too strange to be true? Not at all.

The story of Pluto is a fascinating one. The manner in which the planet came to be found illustrates the chance aspect of scientific discoveries, and the ongoing discussion about whether it is or isn't a planet takes us a long way toward better understanding our solar system.

Finding Pluto

Though Clyde Tombaugh is the person credited with finding Pluto, he was by no means the first person to have looked for it. In 1894 Percival Lowell, at age thirty-nine a Harvard graduate and member of a well-established Massachusetts family, founded an observatory in Flagstaff, Arizona (elevation 7,000 feet), where he spent twenty-three years pursuing his hunch that there was not only water but also life on Mars. He published several books on this topic and also devoted himself to other projects, including the search for Planet X. Lowell had come to believe that the orbits of Uranus and Neptune were being affected by the gravitational pull of another planet. (Later, after Pluto was identified, scientists came to realize that the original calculations regarding Uranus and Neptune were incorrect, and while additional Planet Xes may well be found, there actually were no planet-caused alterations in the orbits of the two previously identified planets.)

While Lowell might not have had an accurate vision of what was happening, his hunch about other planet-like bodies was on the mark. Lowell died in 1916 without ever realizing that his true contribution to science was in his belief of a world beyond Neptune.

How a Farm Boy Came to Discover a Planet

Clyde Tombaugh was about as unlikely a planet discoverer as you can imagine. Tombaugh was born in 1906 in Streator, Illinois, and grew up on a farm near Burdett, Kansas, where he used parts of old cars and farm machinery to create his own telescope, a 9-inch Newtonian reflector, which he completed by 1927. Using the telescope and the dark night skies in western Kansas, he made drawings of Mars and Jupiter and submitted them to scientists at the Lowell Observatory, hoping to get feedback on his drawings. They must have seen something remarkable in the young fellow's drawings as they hired him to come to work at Lowell, even though he had only a high school diploma.

Though Lowell had died in 1916, the observatory was still in pursuit of his dream of finding Planet X, and originally Tombaugh was assigned to use a new 13-inch f/5 camera to photograph the skies in search of this possible planet. Soon he was also given the job of scanning photographic plates using a Zeiss blink comparator, a precomputer device that permitted astronomers to find differences between photographic plates taken of the same area of the sky at different times. (Scanning photographic plates was a very necessary and important job, but it was tedious and was often performed by clerical workers. In those days, it was the type of job women were sometimes hired to do.)

On February 25, 1930, Tombaugh was looking at the plates photographed on January 23 and January 29 when he noticed a speck of light shifting positions on successive plates exactly as a trans-Neptunian planet should. He thought he had found a planet. Though scientists today are encouraged to re-

Percival Lowell.
Source: Library of Congress.

lease information quickly so the rest of the community can start building from any newly discovered base, in those days, confirming discoveries took longer. The scientists at Lowell Observatory did not announce the discovery until March 13, 1930.

The excitement of this discovery—the first planet to be identified since 1846 and the discovery of Neptune—was great. Percival Lowell's dream was realized, right at the observatory he founded, and Americans were enchanted with the thought that the planet had been discovered by a U.S. citizen. Tombaugh was honored by organizations throughout the world. Eventually, he went back to school for his bachelor's and master's degrees and continued to work, discovering a comet, a nova, five open clusters, a globular cluster, and a supercluster of galaxies.

In 1934 the *New York Times* reported that observatories throughout the nation were belatedly identifying images of

Pluto that were taken long before its discovery in 1930. Ironically, one such photographic plate was found at the Lowell Observatory; the photo was taken in 1915, a year prior to Lowell's death. Had the plate been scanned and the planet identified during Lowell's lifetime, the great believer would have been able to overcome the professional derision that his quest caused. He died never knowing that he had already recorded pictures of the solar system's ninth planet.

The Naming of the Newest Planet

The tradition of naming the planets after Greek or Roman gods dates back to ancient times. The original planets, of course, were known to the Greeks and Romans, for they noted that in the night sky some "lights" moved while others didn't, and they called the ones that did *planets* from a Greek term meaning "wanderer."

Today the official naming organization for planets is the International Astronomers Union, but before they had an opportunity to meet after the finding of Pluto, the newspapers were filled with suggestions from all types of people and organizations. Among the names floated out were Atlas, Zymal, Artemis, Perseus, Vulcan, Tantalus, and Idana. Minerva was one possibility that was suggested by the staff at the Lowell Observatory, but a *New York Times* article of March 26, 1930, quotes a Museum of Natural History scientist who reminded scientists that asteroid #93 (out of about 1,000 that were then identified and named) was named Minerva.

The name Pluto was actually suggested by an eleven-year-old girl who lived in Oxford, England. This wasn't just any little girl, however. Venetia Burney's grandfather, Falconer Madan, was librarian of the Bodleian Library of Oxford University, and he happened to read the story of the new planet's discovery at breakfast on March 14, 1930, and mentioned it to his granddaughter. Venetia's class had just been on a nature walk where they had laid out the planets to scale—she knew from the nature walk that Saturn was a shocking 1,019 paces away from Earth, so she was well aware of how very far away a planet beyond dark Neptune would be. (Currently Neptune is actually farther away than Pluto because of the elliptical nature of the orbits.) In addition, she had been reading a book called *The Age of Fable*, so she was familiar with mythology. As a result, she suggested to her grandfather that Pluto, the name of the god of the underworld, would be fitting for the newest planet. Her grandfather dutifully contacted one of his associates, a former astronomer royal, and in turn, Herbert H. Turner sent a telegram on to the Lowell Observatory.

The combination of the fitting mythological name and the fact that the first two letters of the name could serve to honor the initials of the original astronomer with the dream, Percival Lowell, meant the name Pluto stuck.

Six Facts about Pluto

- **Pluto is the smallest planet in the solar system, smaller than Earth's moon, and half the width of Jupiter's moon, Ganymede, which is larger than Mercury and the largest moon in the solar system.**
- **Pluto orbits the sun on a different plane from that of the other eight planets.**
- **While all planets have elliptical orbits, Pluto's orbit is more elliptical than that of any of the other planets, meaning that it can come closer to the sun than Neptune, but then it travels almost 2 billion kilometers farther away from Neptune's orbit.**
- **Pluto has one moon, Charon, that is not much smaller than Pluto and was identified in 1978. Then in 2006, there was more news: Astronomers using NASA's Hubble Space Telescope announced that they had spotted two more small moons circling Pluto. It is thought that all three moons formed in the aftermath of a collision between a large meteor and Pluto, and scientists think that eventually even more moons may be found.**
- **Pluto's journey around the sun takes 247.7 Earth years. This means that it still has a good number of years left before it has completed a full orbit since its discovery in 1930.**
- **A day on Pluto lasts for 6 days and 9 hours, meaning that it has the second slowest speed of rotation in the solar system (after Venus, which takes 243 days to turn on its axis).**

So Is It or Isn't It?

For a good number of years, life with nine planets felt right to the world. By 1931 the Disney character Mickey Mouse had acquired a pup named Pluto; Clyde Tombaugh raked in numerous honors and went on to found an astronomy department at New Mexico State University; and school children learned the names of all nine planets.

Then in 1996, trouble began to brew when a group of scientists put forth the thought that perhaps Pluto wasn't actually a planet. At the time it was discovered, Pluto was the only known object beyond Neptune in the solar system. (When Pluto's first moon, Charon, was spotted, that seemingly confirmed Pluto's planet status.) Today astronomers have found about 1,000 other small icy objects (and there may be many more) beyond Neptune rotating around the sun, in an area called the Kuiper Belt. To complicate matters, there is disagreement over the exact definition of a planet.

To continue this discussion, we need to step back for a moment and consider what constitutes a planet. To the first astronomers—the Greeks and Romans—the planets they could see (Mercury, Venus, Mars, Jupiter, and Saturn) were viewed as "stars" that were brighter than any other planetary bodies, and they traversed the heavens in predictable ways.

Since that time, scientists have tightened up their definitions of what a planet is (though it is still a subject under discussion). In general, it is agreed that a planet is a large, spherical natural object that directly orbits a star and does not generate heat by nuclear fusion the way a star does. This definition also distinguishes planets from asteroids and comets. Some scientists feel that any object in the solar system large enough that gravity has shaped it into a sphere should be called a planet. But that would also add quite a few asteroids and other Kuiper Belt objects. Another possibility is to arbitrarily call anything larger than Pluto a planet.

Scientists, however, are also in agreement about another thing: Pluto is not your ordinary planet. The other eight planets fall into much more orderly categories.

COMETS: Comets are generally quite small—under 10 kilometers or so—and when they get close enough to the sun they release water from their surface; the moisture combines with dust or dirt to create huge tails in the sky. Each comet actually has two tails—one of gas and one of dust.

ASTEROIDS: The majority of asteroids are small, irregularly shaped objects with airless surfaces. They are usually located in the asteroid belt between Mars and Jupiter; however, they can be found as near as Venus or as far out as beyond Pluto.

SATELLITE: This is an object like our moon that orbits another larger object, not neccessarily the sun.

Asteroids and comets directly orbit a star (our sun) but generally are not large enough to pull themselves into a spherical shape by gravity.

The inner four planets, Mercury, Venus, Earth, and Mars, are all classified as terrestrial planets because their surfaces are rocky. The Jovian planets, Jupiter, Saturn, Uranus, and Neptune, are known for being gas giants. And then there's little Pluto. . . .

Not only is Pluto the smallest of the planets, but it is even smaller than Earth's moon and just about half the size of the next smallest planet, Mercury. On a bad day, scientists refer to it as "a big ice ball," which actually is an accurate description of it. It has an icy composition very much like that of a comet, though it is 1,000 times the size of a normal comet. It also orbits the sun at a 17-degree tilt relative to the plane in which the other planets move. It is believed to be composed mostly of water ice with a relatively thin layer of nitrogen ice mixed with small fractions of methane and carbon monoxide.

The Fight Goes On

Toward the end of the twentieth century, the American Museum of Natural History was completing the work on the remodel of the Hayden Planetarium and the related display space, now named the Rose Center for Earth and Space. When the exhibit area opened and revealed that Pluto was listed not as a planet but as a Kuiper Belt object, Disney might as well have announced that Pluto the Pup was to be euthanized, so great was the outcry. In addition to hearing from scientists, the museum was flooded with plaintive letters from schoolchildren begging that Pluto be reinstated as a planet.

As a result of the uproar, the International Astronomers Union, the organization responsible for naming and classifying everything in space, agreed to create a committee to study how to create a "proper" definition of a planet. Defined one way, Pluto would be "in," as might be a good number of additional objects that are known or that may be discovered; defined another way, Pluto might be "out," and we would be left with just eight planets.

Another Discovery Affects the Picture

In July of 2005, the scientific world hit another speed bump. Spanish astronomers came across information on the Internet about work by scientists at the California Institute of Technology, and the Spanish discovery pushed into print the announcement by Michael Brown of a new discovery—a "planet" very much like Pluto—only bigger. Officially known as 2003 UB313, but nicknamed Xena "because it's easier to refer to that way," writes Mike Brown ("A Tenth Planet beyond Pluto!" www.gps.caltech.edu/~mbrown/), the new discovery is also a member of the group of icy objects that make up the Kuiper Belt. In October 2005 the discovery of a moon that orbits this new body intensified the discussion.

As of 2006, scientists were able to announce its size and have determined that it has a diameter of approximatiely 1,860 miles. (Pluto is only 1,400 miles in diameter.)

The next step? The International Astronomers Union (IAU) must come up with a definition for planet so that Mike Brown's objects can be named according to their classification.

Here are some possibilities for what may be decided:

1. The IAU may decide to define planets more narrowly, excluding 2003 UB313, Pluto, and the "plutinos," as some of the other Kuiper Belt objects are referred to.

2. The IAU may decide that precedent prevails, that seventy-five years of schoolchildren falling in love with "little Pluto" has given Pluto and other similar ice balls the right to planetary status.
3. They may decide to classify planets: the terrestial planets (Mercury, Venus, Earth, and Mars) and the Jovian planets (Jupiter, Saturn, Uranus, and Neptune). A third term (ice planets? the frosties?) would cover Pluto, UB313, and all else that follows.

So stay tuned. The reason this isn't in your science book is because it's still happening!

Pluto Probe Is Launched

On January 19, 2006, an unmanned NASA spacecraft blasted off on a mission to Pluto, a nine-year, 3-billion-mile journey. It is the fastest spacecraft ever launched, capable of reaching 36,000 miles per hour, but it will still take nine and a half years to reach Pluto and the outer reaches of the solar system.

The New Horizons launch is planned so that by 2007 the rocket will be near enough to Jupiter that Jupiter's powerful gravitational field will slingshot it on the way to Pluto. It will not land on Pluto but will photograph it, analyze its atmosphere, and send data back across the solar system to Earth.

Then the mission is to continue past Pluto, possibly visiting large objects in the Kuiper Belt, which contains comets and small planets. Scientists are hopeful that this will provide helpful information about how the sun and planets formed.

In tribute to Pluto's discoverer, New Horizons contains some of Tombaugh's ashes. His ninety-three-year-old widow was able to watch the liftoff from about four miles away.

Other Excitement in Space

In early 2006, scientists were beside themselves with excitement upon the safe return of the Stardust spacecraft, which

Litter Is Litter in Space, Too

NASA scientists report that as of 2006, more than 9,000 pieces of space debris are orbiting the earth, a hazard that can only be expected to get worse in the next few years.

The most debris-crowded area is between 550 miles and 625 miles above the earth, and while the junk is not a great risk factor for manned space flight, it can pose a risk to commercial and research flights that travel beyond the area where the space shuttles normally travel.

At this point there is no viable solution, technically or economically, for removing objects from space. However, scientists will need to get to work on answers as the problem is only going to get worse.

made a safe landing in Utah after a seven-year 3-billion-mile journey into outer space.

The reason for the excitement is that its cargo—an estimated million particles of comet dust along with perhaps 200 grains of stardust—is expected to provide information about how the solar system and earth formed—and ultimately, something about the origins of life on earth, since the composition of these particles has been essentially unchanged for 4.5 billion years.

The research on these particles is expected to take decades and will be farmed out to 160 scientists around the world. In addition, a selection of volunteers (65,000 people signed up to help) will use software provided by NASA to work at their home computers, scanning images of portions of the collector cells, in an effort to more quickly locate the tiny particles to be studied. (Most of these particles are less than one-tenth the width of a human hair.) The grains are believed to be the pristine remains of the birth of the solar system some 4.6 billion years ago.

Space Travel for Regular People

While you can't exactly phone a NASA reservation agent yet, the government is beginning to look toward a day when space tourism is available to ordinary people with big bank accounts. (In 2001 American businessman Dennis Tito paid Russian space chiefs a reported $20 million to travel on one of their space ships, thus becoming the first space tourist.)

The FAA is currently considering rules for training and medical qualifications, and the full proposal is on line at ast.faa.gov/files/pdf/Human_Space_Flight_NPRM.pdf.

PART FOUR

Protecting Our Living Planet

15

Little-Sung Heroes
The Environmental Crusaders

For most areas of science, a specialized education is necessary to move forward in the field, but when it comes to being an environmentalist, the main requirements are an appreciation for the world around us and a willingness to take steps to save it. As this chapter demonstrates, what started as a simple appreciation for nature has become a fight to keep progress and nature in balance.

To find a beginning of the American environmental movement, we need to look back to Henry David Thoreau, one of the most influential figures in American thought and literature.

Henry David Thoreau (1817–1862)

When we envision life 150–175 years ago, we tend to think of people living simply on farms or in small towns; those who lived in cities would have gone about their days to the rhythm of the gentle sounds of horse-drawn carriages and would have been spared the noxious exhaust fumes from diesel trucks or buses that we routinely breathe today—or so we imagine.

But though life may have been a bit simpler, environmental awareness was low—our country's resources seemed limitless, and landowners pushed for developing anything that promised profit. Coal smoke puffed blackly from factory stacks, and wood-burning home fireplaces fouled the air as well. No one cleaned up after their horses, so pedestrians in towns and cities had to tread carefully. Because garbage was often disposed of by tossing it out a window, the stench on a city street, particularly in the heat of the summer, was often unbearable.

So when Thoreau moved to Walden Pond for his "experi-

The site of Henry David Thoreau's hut near Walden Pond in Massachusetts. Source: Library of Congress.

ment in living," it was something different—even for that time. And while his writings did not sell well during his lifetime, his message in *Walden; Or, Life in the Woods* (1854)—that harmony with nature was possible when one lived simply, reading, writing, and walking in the woods—came to public attention later on.

In 1862, the year of Thoreau's death, the book was brought back into print. Since that time, *Walden* has become one of the most widely read and influential books throughout the world, conveying to people everywhere that Thoreau's message of living in harmony with nature is still vital.

John Muir (1838–1914)

Shortly after Thoreau, a fellow named John Muir took an interest in improving our environment and was to become one of the country's most influential conservationists.

Muir was born and raised in Dunbar, Scotland. When he was eleven, his family emigrated to the United States and settled near Portage, Wisconsin. Once Muir was old enough to work, he supported himself clearing forest land. In 1867, however, a terrifying machinery accident took the sight in one of his eyes and changed the course of his life. After the incident, Muir resolved to immerse himself in all that he might have missed if he had become totally blind.

Muir's first great wilderness adventure was a thousand-mile walk from Louisville, Kentucky, to Savannah, Georgia, at a time when the area was sparsely settled and there were no asphalt roads. It made a big impression on him and provided a lesson he became intent on sharing with others—that nature must be respected.

In 1868 he journeyed to San Francisco and then on to the Yosemite area, where he explored the Sierra Nevada Mountains and first saw breathtaking cliffs with waterfalls that tumbled hundreds of feet and immense trees with trunks that were more than 100 feet around. Eventually he married and moved to Martinez, California, but he continued to travel to Yosemite, where he observed the devastation of the mountain meadows and forests by sheep and cattle and witnessed the changes that were already happening because of the logging industry. He became concerned that unless something was done, the wilderness he loved would disappear.

Muir soon became Yosemite's most outspoken crusader, but his position as defender of the wilderness was not a popular one in the nineteenth century. This was a time when settlers were thirsting to acquire more land and thought nothing of exploiting it through farming, logging, damming streams, and mining. Even public lands were made available for exploitation—in 1872, Congress passed the now-infamous Mining Law, under which companies and individuals could buy the mining rights for public land thought to contain minerals.

Muir remained intent on his cause, and eventually his writings came to the attention of Robert Underwood Johnson, the editor of *Century*, one of the most prominent magazines in the country at that time. One of the first of Muir's articles to be published in *Century* advocated that a national park should be created for the Yosemite Valley. Johnson not only published

Teddy Roosevelt and John Muir at Yosemite. Source: Library of Congress.

the article but lobbied hard for Congress to take action. On June 30, 1864, President Abraham Lincoln signed a bill granting Yosemite Valley and the Mariposa Grove of Giant Sequoias to the State of California as an inalienable public trust. This was the first time in history that the federal government had set aside scenic lands simply to protect them and to allow for their enjoyment by all people.

Though Yellowstone became the first official national park in 1872, by 1890 Muir had won his campaign for Yosemite to be so designated. National parks were a new concept, and the federal government did not yet have a clear view of how the parks could best serve public interests. Johnson and others suggested to Muir that an association be formed to protect the newly created Yosemite National Park from stockmen, who did not respect park boundaries. In 1892, Muir and a number of his supporters founded the Sierra Club to, in Muir's words, "do something for wildness and make the mountains glad." Muir served as the club's president until his death in 1914.

In 1901, Muir published *Our National Parks,* which brought him to the attention of President Theodore Roosevelt, who visited Muir in Yosemite in 1903. Discussions between the two men laid the foundation for Roosevelt's innovative and notable conservation programs, including establishing Yosemite National Park and the first national monuments by presidential proclamation.

Unfortunately, Muir lost his last conservation battle—to save the Hetch Hetchy Valley. The city of San Francisco wanted to dam the Tuolumne River, which runs through Yosemite, to create a source for drinking water and hydroelectric power. In 1913, Congress passed the Raker Act, authorizing the construction of O'Shaughnessy Dam, and soon the beautiful valley was flooded to make way for progress. Ironically, there is now a movement afoot to restore Hetch Hetchy.

Teddy Roosevelt (1858–1919)

In photographs or sketches, Teddy Roosevelt, our twenty-sixth president (1901–1909), is most frequently depicted on horseback as he would have appeared as part of his Rough Riders group, or in a pith helmet and khakis, preparing for one of his big game–hunting expeditions. To do justice to Roosevelt's memory, we need to put these images in the context of the day, for Roosevelt was not the exploiter he might seem to have been and is actually the president who deserves credit for protecting much of the American landscape.

Though today hunting would be inconsistent with being an ardent environmentalist, at the beginning of the twentieth century there was little understanding of nature as a limited resource. Hunting was also a method of collecting and learning. While there is no doubt that Roosevelt relished the actual hunt, his trips were sometimes to gather skins and skeletons for American museums—his bounty found homes at both the Smithsonian Institution in Washington and the American Museum of Natural History in New York City.

In all likelihood, Roosevelt's love of nature can be attributed to the circumstances of his boyhood. Growing up, he suffered greatly from asthma, and he combated his ill health with exercise, often in the form of long walks. Using the out of doors as a "cure" eventually became a part of Roosevelt's personality. Throughout his life, Roosevelt frequently sought refuge in the outdoors. When his first wife and his mother died on the same day in 1884, Roosevelt responded by escaping to a ranch he owned in the Badlands of Dakota Territory. And during his term as vice president, he was on a hiking trip with his family in the Adirondacks when he received word that McKinley had been shot and that he was to become president. A trip down the Amazon was part of the solution to his diminishing postpresidential political power.

As a result of his love of being outdoors, Roosevelt felt that our resources should be protected, used in moderation, developed when necessary, and passed on to future generations. Roosevelt was the first to introduce the concept that land, water, minerals, and forests of the country needed to be held in trust by the government—not sold or given away to the highest bidder as had been the practice up until that time. He added enormously to the national forests in the West, reserved lands for public use, and fostered large-scale irrigation projects.

Toward the end of his term, Roosevelt took an important step that furthered his conservation agenda. He appointed a National Conservation Commission in 1908, which undertook creating the first inventory of the country's natural resources.

Marjory Stoneman Douglas (1890–1998)

In earlier days, people didn't grasp that tampering with nature could cause problems. As a matter of fact, in 1850, in anticipation of future development nationally, Congress passed the Arkansas Bill, or the Federal Swamp and Overflow Act, which provided for the title of wetlands to be transferred from the federal government to the states. Florida was ready for this, and state officials arranged to drain and reclaim parts of the Everglades, a unique subtropical marshland in the southern part of the state that extends from Lake Okeechobee on the north to Florida Bay on the south. Large tracts of land were drained in preparation for agricultural development, but only lands immediately bordering Lake Okeechobee were actually farmed at that time. (The first thorough studies of the Ever-

Why What Happens Elsewhere Makes a Difference

One might well wonder, "Why should I care what happens with Arctic drilling or whether the Everglades are saved? I live in Texas [or insert your state]."

The economic abundance of the modern world depends on the health of its air, soil, and water. In an interview, an environmental attorney explains it this way:

> **The world is one big ecosystem, and we have to protect it and keep it in balance. If man uses up all of one of earth's resources, he will find or create another, but if progress alters the balance of nature to the point that extended land areas—say all of the U.S. southwest or big stretches in India—have no rainfall, it will impede our ability to grow crops, feed people, and even to have water to drink. That would be a natural disaster that could have a very bad ending.**

And if we don't pay attention to the environment near our homes, we will continue to see headlines like the one that appeared in Westchester County, New York, in late 2005: "Contamination worse than thought at fields in Valhalla." As a result of contaminated soil, the high school administrators feared that they would not be able to use the ball fields that spring.

Today we know the dangers of overdevelopment. We understand that too much asphalt covering our land leads to pollutant-laden runoff, and often, to flooding; that overbuilding can mean destruction of the habitats of birds and animals that once populated our land; and that tampering too much with nature can lead to disaster. Think of what Hurricane Katrina did to New Orleans. While the creation of levees usually prevents flooding, the levees also prevent the deposit of sediment that replenishes the land. As a result of these human-made creations in New Orleans, and the cutting of pipelines and shipping channels, the land sunk. When Katrina and its aftermath hit, the destruction by flooding was devastating.

---

glades—conducted after fires burned through the drained land in 1939—concluded that most of the southern part was unfit for cultivation.) In addition, a state-formed Improvement Fund was created and allowed railway companies and developers to buy a great deal of land.

In the 1940s, a person came along who was to forever change the fate of the Everglades. Marjory Stoneman Douglas, a reporter for the *Miami Herald*, was asked to do a book on the

Miami River for a series on American rivers. As she began work, she realized the river was only a part of the story, and she convinced her editor to accept a book that covered all of the Everglades.

Over time, Douglas began to see how the rapid commercial development of southern Florida was threatening the vast, slow-moving stream of shallow water of the Everglades and the saw grass—the "river of grass" as she called it—that provided a safe haven for so many unusual plants and birds and animals. Douglas realized that what was being treated by developers as a "worthless swamp" actually provided an ecologically necessary service that made southern Florida habitable. What the developers didn't understand was that in hoping to drain and develop the swampy area, they would be destroying an important piece in the cycle of nature. Her 1987 autobiography, *Voice of the River* (Sarasota, FL, Pineapple Press, 1987) summarizes the Everglades' role as an important watershed for southern Florida: "Much of the rainfall on which South Florida depends comes from evaporation in the Everglades. The Everglades evaporates, the moisture goes up into the clouds, the clouds are blown to the north, and the rain comes down over the Kissimmee River and Lake Okeechobee."

Douglas's book caught public attention, and others joined her in the cause. Eventually the public outcry attracted the interest of then-president Harry Truman, who put through an executive order later that year (1947) to protect more than 2 million acres (8,000 square kilometers) as Everglades National Park.

Her name became synonymous with the Everglades for her tireless, ground-breaking efforts. To build a base of support for the Everglades, she created the Friends of the Everglades organization in 1970 and devoted much of the rest of her life to protecting the area.

Florida (1994) and the federal government (1996) launched long-term reclamation projects aimed at removing levees, reflooding drained swampland, and otherwise "replumbing" the Everglades. In 2000 Congress passed legislation enabling the multibillion-dollar project. The cost would be

Simple Acts Can Destroy Balance

Sometimes even a seemingly harmless activity can destroy the balance of nature. During the last five years, Burmese pythons imported from Vietnam have become very popular pets. Owning these snakes is legal, but problems arise when they grow up, as the snakes can easily reach up to fifteen feet, no longer a convenient size to keep around the house. As a result, well-meaning pet owners have released them into area swamps, and over a ten-year period, the rangers in the Everglades have discovered a good number of pythons. In 2004 alone, sixty-one of these snakes were found in the park. Unfortunately, the release—and subsequent successful breeding of these snakes—is bad news for birds and animals native to the Everglades. The pythons appear to be eating wood storks and mangrove fox squirrels and are taking away food and habitat from the native species. Park rangers are now using dogs to track down the pythons to protect the ecological balance of the Everglades.

split between the state and the U.S. government. Douglas would undoubtedly be pleased.

Rachel Carson (1907–1964)

Rachel Carson has been called the mother of the modern environmental movement for what she did to awaken the world to the hazards of the indiscriminate use of synthetic chemical pesticides, particularly DDT.

Unlike some of the environmentalists we meet in this chapter, Carson came to this effort with serious science credentials. She obtained an MA in zoology from Johns Hopkins University and soon joined the U.S. Bureau of Fisheries as the writer of a radio show about ocean life. In 1936, she was the first woman to take and pass the civil service test, and the Bureau of Fisheries hired her as a full-time junior biologist. Over the next fifteen years, she rose to become the chief editor of all publications for the U.S. Fish and Wildlife Service. During the 1940s, Carson began to write books on her observations of life under the sea, and by 1952 she was ready to resign her government job to devote her time to writing. Her books (*Under the Sea*

Wind, 1941, *The Sea Around Us*, 1951, and *The Edge of the Sea*, 1955) sold well, earning Carson a reputation as a well-respected naturalist and science writer.

When she received a letter from a friend in Massachusetts telling her of the large number of birds that were dying on Cape Cod as a result of DDT sprayings, Carson wanted to pursue the topic and tried to interest a magazine. Though Carson was a best-selling author, she had no luck at placing an article on what was turning out to be a controversial topic. She decided it had to be the focus of her next book.

While people have been experimenting with pest control for centuries, it was not until 1942 that DDT, the most powerful pesticide the world had ever known, was introduced. Up until this time, most pesticides that were in use were limited to destroying one or two types of insects, but DDT was capable of killing hundreds of different kinds at once. At first, DDT was lauded as a great advance. It distinguished itself during World War II, clearing South Pacific islands of malaria-causing insects for U.S. troops, and in Europe it was being used as an effective delousing powder. The world was so excited about its creation that, ironically when viewed today, its inventor was awarded the Nobel Prize.

Though Carson couldn't sell a magazine article on the topic, she did sell *Silent Spring*, which was published in 1962. It was to become one of the landmark books of the twentieth century. It alerted the general public to the dangers of pesticides, particularly to humans. In it, Carson described how DDT entered the food chain and accumulated in the fatty tissues of animals, including human beings, and caused cancer and genetic damage. According to Carson, a single application on a crop killed targeted insects as well as countless others, and the pesticide remained toxic in the environment even after it was diluted by rainwater, causing irrevocable harm to birds and animals and contaminating the entire world food supply.

The chemical industry was outraged, and Carson's critics called her "Carrie Nation," after the hatchet-wielding temperance advocate. Carson quietly continued to push for new policies to protect human health and the environment. She stressed that people are a vulnerable part of the natural world, and anything used widely to destroy insects can't be good for long-term human exposure. Stating that she was not opposed to the use of chemical sprays, she was taking a stand against their "indiscriminate use" at a time when their effect was not fully known. *Silent Spring* caught the interest of the public, and President John F. Kennedy ordered a President's Science Advisory Committee to examine the issues the book raised.

The chemical industry retaliated that the book was a gross distortion of fact and that the real threat to survival of humanity is the insects that destroy crops and trees. But Carson had prepared carefully and had fifty-five pages of notes and experts who had read and approved the book.

On April 3, 1963, *CBS Reports* presented a program, "The Silent Spring of Rachel Carson." In it Carson said,

> It is the public that is being asked to assume the risks that the insect controllers calculate. The public must decide whether it wishes to continue on the present road, and it can do so only when in full possession of the facts.
>
> We still talk in terms of conquest. We still haven't become mature enough to think of ourselves as only a tiny part of a vast and incredible universe. Man's attitude toward nature is today critically important simply because we have now acquired a fateful power to alter and destroy nature.
>
> But man is a part of nature, and his war against nature is inevitably a war against himself. The rains have become an instrument to bring down from the atmosphere the deadly products of atomic explosions. Water, which is probably our most important natural resource, is now used and reused with incredible recklessness.
>
> Now, I truly believe that we in this generation must come to terms with nature, and I think we're challenged as mankind has never been challenged before to prove our maturity and our mastery, not of nature, but of ourselves.
>
> A spray as indiscriminate as DDT can upset the economy of nature as much as a revolution upsets social economy. Ninety percent of all insects are good, and if they are killed, things go out of kilter right away.

As a result of Carson's work, DDT came under much closer government supervision and was eventually banned. She created a new public awareness that nature was vulnerable to human intervention. Though the public had never been particularly captured by naturalists who talked of the disappearance of wilderness, Carson's points—the contamination of the food chain, cancer, genetic damage, the deaths of entire species—were too frightening to ignore. For the first time, the need to regulate industry to protect the environment became widely accepted. Environmentalism was born.

Julia Butterfly Hill (1974–)

I conclude this chapter with a brief story about Julia Butterfly Hill, who, thus far, is remembered for what she did rather than what she wrote.

In 1997 Hill became a symbol for the environmental movement when she climbed into a giant 180-foot (55-meter) California coast redwood tree to protect the tree and the forest where it had lived for a millennium. Though Hill was not the only environmentalist to climb up into a tree's branches that day, she stayed in her tree, dubbed Luna, for two years (738 days), until the Pacific Lumber/Maxxam Corporation ultimately agreed not to cut down Luna or the surrounding redwoods and guaranteed a three-acre buffer zone.

Julia Butterfly Hill proves that even today one person can make a difference. Whether you worry about mercury in our fish (chapter 17), global warming (chapter 16), or Arctic drilling, there is always a way that you can make a difference.

Ten Simple Steps

According to Canadian scientist David Suzuki, we needn't climb a tree and live there for two years or lead an environmental movement to improve the environment. At his website, www.davidsuzuki.org/WOL/Challenge/, he offers a Nature Challenge in which he puts forth ten simple steps to take, and he says committing to only three of them will make a difference:

HOME—LIVE CLEAN!

1. Find ways to reduce your home heating and electricity use by 10 percent this year.

2. Replace chemical pesticides on your lawn, garden, and houseplants with nontoxic alternatives.

3. Choose an energy-efficient home and appliances. (Appliances are rated as to efficiency.)

FOOD—EAT LOCAL AND LEAN!

4. Choose at least one day a week to eat meat-free meals.

5. Prepare your meals with food from local farmers and producers for one month this year. (If food is trucked in, this leads to an increase in greenhouse gas emissions and other pollutants.)

TRANSPORTATION—GO GREEN!

6. Watch for fuel efficiency when you buy your next car.

7. Walk, bike, carpool, or use mass transit to get to one of your regular destinations each week.

8. If you are moving, choose a home within a 30-minute bike, walk, or transit ride from your daily destinations.

9. Support alternatives to the car. Contact media and government urging improved public transit and bike paths.

GET INVOLVED, STAY INFORMED!

10. Learn more about conserving nature and share what you've learned with family and friends.

16

It's Getting Hot in Here

The Reality of Global Warming

As the twenty-first century gets underway, one of the hotly contested issues of the day is global warming. Two facts, however, are not disputed:

1. Our world is getting hotter.
2. The energy-consuming activities of human beings are a large reason for the Great Heat-Up.

While some scientists dispute that global warming is an issue that should cause concern, as recently as 2004, consultants hired by the Pentagon released a national security report that envisions a scenario in which global warming causes massive areas of the world to become uninhabitable and causes major food and water shortages. The report predicts widespread migration and war if this scenario were to actually occur.

So what is really going on with global warming?

What Causes Global Warming?

If you recall anything at all about the history of the planet, then you will recall that there have been several ice ages (periods of time when there is a long-term decrease in the earth's temperature, resulting in an expansion of the continental ice sheets, polar ice sheets, and mountain glaciers), with the most recent one ending 10,000 years ago. So obviously, our planet has experienced some major fluctuations in temperature.

The thing that has scientists concerned is the rapid rate of the warming trend over recent years. Over the past fifty years,

Your Weatherperson Doesn't Know

All parts of the country experience fluctuations in the weather. Whether you live in Florida, Maine, or Southern California, your weatherperson or newscaster has probably had occasion to report on a cold snap for your area and then added, "Well, I don't think we're having any global warming . . ."

While outdoor air temperature is what we—and the weatherperson—all notice when it comes to weather, that's not what the scientists keep track of. They are busy documenting glacier size (and rates of melting), the condition of permafrost (the part of the Arctic land area that has historically remained permanently frozen), overall temperature on earth, and water temperatures.

Based on these measures, what any scientist will tell you is that our entire world is heating up.

the average global temperature has increased at the fastest rate in recorded history, and some feel the trend is accelerating: the three hottest years on record have all occurred since 1998.

Among the scientific measures of greatest concern are these:

- Nearly every major glacier in the world is shrinking.
- The oceans are becoming warmer, and as a result, more acidic.
- The loss of sea ice since the late 1970s is equal to the size of Texas and Arizona combined.
- The Arctic's perennial polar ice cap is shrinking at the rate of 9 percent per decade. In September of 2005, scientists reported that the cap is now the smallest it has been in a century of record keeping.
- There is a diminishing difference between daytime and nighttime temperatures.
- Plants are blooming earlier in some areas, and animals' ranges are shifting. The Inuit language has never had a word for "robin" because these birds never made it so far north. In recent years, robins have been seen in this northern part of the world.

Experts feel that the world is now warmer than it has been at any point in the last two millennia, and if the current trend continues, it will likely be hotter than at any point in the last two million years. An additional risk of these rising temperatures is that under the right conditions, even organic material that has been frozen for millennia can break down, giving off additional carbon dioxide or methane and increasing the greenhouse gases even more.

So What?

In 2002, Colorado, Arizona, and Oregon endured their worst wildfire season ever. That same year, drought created severe dust storms in Montana, Colorado, and Kansas, and floods caused hundreds of millions of dollars in damage in Texas, Montana, and North Dakota. According to statistics from the National Resources Defense Council, since the early 1950s, snow accumulation has declined 60 percent in some areas of the Cascade Range in Oregon and Washington.

In addition, scientists make these predictions:

- Melting glaciers, early snowmelt, and severe droughts will cause more dramatic water shortages in the American West.
- Rising sea levels will lead to coastal flooding.
- Warmer sea surface temperatures will fuel more intense hurricanes along the southeastern Atlantic and Gulf coasts. Global warming doesn't create hurricanes, but it does make them stronger and more dangerous. When the ocean becomes warmer, tropical storms can pick up more energy and become more powerful.
- Forests, farms, and cities will face troublesome new pests and more mosquito-borne diseases because insects can migrate into areas that were formerly inhospitable to them.
- Disruption of habitats such as coral reefs and alpine meadows could drive many plants and animal species to extinction.

And this information reflects only what is happening in the United States.

The Carbonation of the Atmosphere

Last year, national and international panels of hundreds of climate experts agreed that most of the warming during the past fifty years has probably been caused by human activities creating carbon dioxide, a gas that traps heat in the air like a greenhouse roof.

Carbon dioxide is the byproduct of so much of what we do, and it is being identified as a major contributing factor in global warming. While much of it comes from the burning of coal and oil at an industrial level, carbon dioxide also comes from driving a car, running a refrigerator, and cooking. These activities all burn fossil fuels, and they all add carbon dioxide to the atmosphere, creating a greenhouse effect, and this results in a warmer atmosphere. (Other substances such as methane also contribute to the warming effect.)

In 1979 the National Academy of Sciences undertook its first rigorous study of global warming. Climate modeling—the method used to project the outcome if certain climate changes occur—was in its infancy at that time, but even then, the results of the work done on what carbon dioxide was doing to our atmosphere was alarming enough that President Jimmy Carter called on the academy to investigate. As a result, an ad hoc study group on carbon dioxide and climate was formed.

The American Geophysical Union, one of the nation's largest and most respected scientific organizations, has officially taken a stand, commenting that "natural influences cannot explain the rapid increase in global near-surface temperatures."

What is frightening is the speed at which we have increased carbon dioxide levels. In the 1780s (around the time of the Revolutionary War), carbon dioxide levels stood at about 280 parts per million. (This was about what they were 2,000 years before that, too.) The Industrial Revolution began to drive up levels of carbon dioxide—gradually at first. It took almost 150 years to get to 315 parts per million. By the 1970s, the levels had reached 330 parts per million, and then by the 1990s, they hit 360 parts per million.

For all practical purposes, this effect is irreversible. While it is possible to increase CO_2 levels relatively quickly, it is very difficult to bring them down again. Carbon dioxide is a persistent gas that hangs around for about a century.

While it may not be easy, we need to get started, or this will be a nightmare that will affect our children—and our children's children.

What to Do

Though Americans make up only 4 percent of the world's population, the United States produces 25 percent of the carbon dioxide from fuel burning—by far the largest share of any country, and currently more than China, India, and Japan combined. But these figures will change as other nations grow.

And while the United States is by far a major contributor to the problem of global warming, the issue requires worldwide support. The Kyoto Protocol is an international agreement intended to require countries to reduce the emissions of heat-trapping gases that lead to global warming and climate change. While the United States has agreed to continue talks, the American delegation has argued for voluntary cutbacks (instead of signing the agreement that would guarantee cutbacks by all who signed) until there is a way to assure that China and India—two countries who are adding coal-burning factories at an almost exponential pace—will toe the line as well. While this stance is somewhat understandable, some would say that the United States is missing an opportunity to lead the way to a world that shows it can solve its problems.

In the meantime, scientists say that unless we curb global warming emissions, average U.S. temperatures could be 3–9 degrees higher by the end of the century. Here are some ways we could begin:

- Reduce pollution from vehicles and power plants. The United States already has the technology for reducing the gas emissions from factories and cars. Unfortunately, car manufacturers and industry in general have put pressure on Congress to halt or delay the enforcement of rules that would create these changes. The popularity of SUVs over the last fifteen years has resulted in a 20 percent increase in transportation-related carbon dioxide since the early 1990s.
- Phase out the old coal-burning power plants and replace them with cleaner plants, and we certainly should stop subsidizing coal-burning factories. (For more reasons to do this, refer to chapter 17.)
- Turn more toward renewable energy. California has required its largest utilities to get 20 percent of their electricity from renewable sources by 2017, and New York has pledged to compel power companies to provide 25 percent of the state's electricity from renewable sources by 2013.
- Create stricter efficiency for electrical appliances. A Clinton-era change that resulted in a 30 percent tighter standard on home central air conditioners and heat pumps has the net effect of preventing the emission of 51 million metric tons of carbon. This is like taking 34 million cars off the road for a year. A household that uses more energy-efficient appliances (as noted by the Energy Star label), including heating and cooling equipment, kitchen appliances, and computers, could prevent the release of 70,000 pounds of carbon dioxide over the lifetime of the products. That pollution savings is equivalent to taking a car off the road for eight years.
- Conserve energy in small ways as well. Fluorescent lightbulbs are better than incandescent; a fluorescent bulb can keep nearly 700 pounds of carbon dioxide out of the air over the bulb's lifetime.
- Make a greater investment in environmental capital such as forests, mangroves, coral reefs, and other natural resources that are valuable not only as resources but also as protection of the overall eco-health of our world.

- Stop the destruction of the Amazon rain forests. Trees pull carbon out of the air and store it in their tissues, thereby helping to cancel the effect of burning fossil fuels. Dead trees return carbon to the atmosphere, contributing to the problem.
- Plant more regular forests—see above.

And If You Ever Think Your Voice Doesn't Count . . .

In 1997, the Senate had occasion to vote on a global warming policy—whether or not to implement limits on heat-trapping pollution—and not one senator voted in favor of limits. By 2003, when Senators John McCain and Joseph Lieberman introduced legislation to cap and reduce global warming emissions, the Senate came just a few votes shy of passing the measure. We all need to let our congresspeople know that this issue is important to us.

The global warming issue is one with "no do-overs," a message many scientists worldwide would like governments, industry, and individuals the world over to hear.

17

Mercury

From Early Medicine to Environmental Scourge

In chemistry class, you learned about mercury as a metallic element (atomic number 80; chemical symbol Hg) that is probably best known for its use in thermometers, though this use is declining due to the current ban on nonprescription mercury fever thermometers in a number of states. You may also know that mercury is a heavy, silvery, transition metal (refers to their transitional position in the periodic table of elements), that it is sometimes referred to as *quicksilver*, and that it is one of the five elements that are liquid at or near room temperature.

But mercury has a long and varied history, and its story goes well beyond thermometers and the chemistry lab. For thousands of years, mercury has been used by people all over the world in many different ways, and only recently have we realized the terrible danger this element poses to all of humankind.

The Attraction of Mercury

Spain produces about 60 percent of the world's supply of mercury, and certainly as recently as the first half of the twentieth century this was very much a point of national pride.

Mercury is mined from a red ore in the Almadén region, and an interesting and illustrative story comes out of medieval Spain. Because it was easily available, and so pleasing to the eye, the occupants of the Spanish palaces ordered that mercury pools should be used for ornamentation. The reflective silver created a beautiful, dramatic effect, but it is reported that people who lived in those palaces were prone to illness, suffering symptoms such as tremors, drooling, and paranoia—now known to be signs of mercury poisoning.

The Spanish must not have understood the cause and effect involved with mercury, because as recently as 1937, Spain commissioned sculptor Alexander Calder to create a mercury fountain for the Spanish Pavilion at the World's Fair in Paris. Calder's mercury fountain was in the entrance hall, opposite Picasso's *Guernica*, both designed specifically for the exhibition and intended as a political statement during the then-ongoing Spanish Civil War. They were created to protest Ferdinand Franco's siege of the Almadén mercury mines, which were a major symbol of national pride at the time.

Mercury Fountain was Calder's first major commission, and today the fountain is permanently installed at the Fundacio Miro in Barcelona; however, it is now housed behind glass.

What It Actually Is

Mercury is found both in its elemental state and in organic and inorganic compounds, and it is present in the environment as a result of human activity as well as from natural sources such as volcanoes and forest fires. It has been used for three thousand years both in medicine and industry. There are three main forms of mercury:

1. Elemental mercury is the silvery liquid used in thermostats, barometers, batteries, and folk medicine, and by experimenting students in chemistry class. It vaporizes quickly when heated and is toxic in its vaporized form.
2. Inorganic mercury salts are used both in antiseptic creams and ointments and in electrochemistry.
3. Organic mercury compounds such as dimethylmercury or methylmercury are created when mercury in the air lands in water—or on land and is washed into the water—where bacteria can change it into methylmercury, a highly toxic form of the substance that builds up in fish, shellfish, and animals that eat fish. These are by far the most dangerous

> forms of mercury as they can be readily absorbed by the body.

In America, the use of mercury caught on during Gold Rush days when men found it could be used in the amalgamation of gold and silver. Then as the Industrial Revolution got under way, people found all types of uses for mercury, including the creation of daguerreotypes, the silvering of mirrors, and as a preservative. As a matter of fact, because of its preservative qualities it was a valuable ingredient in our wall paints up until 1990, when its use was outlawed.

Mercury is also used in the manufacture of industrial chemicals or for electronic and electrical applications. Gaseous mercury has been used in some forms of neon sign advertising, and liquid mercury is sometimes used as a coolant for nuclear reactors.

Mercury has a corrosive effect when applied to metal, and it is said that the Allies sent soldiers deep into enemy territory to sabotage German planes during World War II by applying a mercury paste to the micron-thin layer of aluminum that holds a plane together. If the story is true, this technique would have been very effective as the mercury would have caused the metal to rapidly corrode, and the planes would have fallen apart.

Mercury's Use as a Medicine

Mercury in the form of its natural reddish pigment, known as cinnabar, was used for cosmetic purposes by ancient Egyptians and Chinese, and the Greeks created medicinal uses for it, which continued throughout history. From the sixteenth century right up until the mid-twentieth century, mercury salts were a prime treatment for syphilis.

Both Presidents Andrew Jackson and Abraham Lincoln are known to have taken mercury. In the nineteenth century, calomel (mercurous chloride) was considered a cure-all and was taken by Andrew Jackson. Because locks of his hair have been preserved, we have been able to test them. (Jackson received frequent requests from the public for samples of his hair, and because he accommodated these requests—and the samples were preserved—we have this evidence for study.) Researchers analyzed samples from 1815 and 1839, and both showed high levels of mercury. Historians report that many of Jackson's physical problems—digestive ailments, excessive salivation, headaches, hand tremors, and dysentery—may have actually been caused by this mercury "treatment."

Recent accounts have been written about Lincoln's bouts of depression and have reported on his use of the "blue mass" pill. The "blue mass" was a popular treatment used in the 1800s for numerous conditions including depression, constipation, toothaches, and child bearing. Elemental mercury was its main ingredient, combined with licorice root, rose water, honey, sugar, and a confection of rose petals. Researchers now know that these pills contained so much elemental mercury that they may have caused mood swings, tremors, and neurological damage in Lincoln. (It is thought that the typical pill contained 9,000 times more mercury than is safe.) Fortunately for history, Lincoln stopped taking these blue pills shortly after his inauguration.

Early in the twentieth century mercury was being given to children annually as a dewormer, and it was sometimes used in teething powder for infants. Even now mercury compounds are found in some over-the-counter medications such as topical antiseptics, stimulant laxatives, diaper rash ointment, eye drops, and nose sprays.

And if you are over fifteen or twenty years old, then you probably remember your mom dabbing mercurochrome—that red stuff—on your cuts and scrapes. Mercurochrome predated federal oversight on medications, and the government did not focus on older topical antiseptic treatments until the 1970s. They began a review of mercurochrome in 1982, but not until 1998 did the FDA finally pronounce that it was "not generally recognized as safe and effective" as an over-the-counter antiseptic and prohibited its sale across state lines.

The most recent issue concerning mercury and medicine has been its use as a preservative in childhood vaccines. (It's also used this way in tattoo inks.) At the same time that the

"Mad as a Hatter"

The phrase "mad as a hatter" derives from the use of mercury. From the mid-1700s on, a process called *carroting* was used to make felt hats. Animal skins were rinsed in an orange solution consisting of a mercury compound, and the process separated the fur from the pelt and helped make it possible for the fur to be matted together. The solution and the vapors were highly toxic, and symptoms among hatters included tremors, severe mood swings, insomnia, dementia, and hallucinations. The United States banned the use of mercury in the felt industry in 1941.

number of required vaccines has increased, there has been an increase in the diagnosis of autism in children. Though studies have not proven a link between thimerosol, the mercury preservative that has been used in vaccines, and autism—scientists do not yet know why we are seeing more autism—there is agreement that mercury is a neurotoxin that can cause great harm to a developing nervous system. As a result, the National Resources Defense Council and others successfully pressed for the removal of thimerosal from childhood vaccines. Now all vaccines are available in a mercury-free form, though some flu vaccines still contain it as a preservative, and some forms of the diphtheria and tetanus vaccines have trace residues.

Dentistry

But even if we avoid mercury in the medicines of today, a visit to the dentist may reveal that you have it in your mouth. If you have any "silver" fillings, then your mouth contains mercury; elemental mercury is the main ingredient in dental amalgams.

The discussion about the safety of filling teeth with a mercury-containing substance has been a long one. As early as 1843, the American Society of Dental Surgeons required members to sign a pledge that they wouldn't use it; however, it is still sometimes used today. Although some health activists claim the mercury leaches out of the fillings and into the body,

the Food and Drug Administration statement of 2002 reaffirmed the mainstream view: "No valid scientific evidence has shown that amalgams cause harm to patients with dental restorations, except in the rare case of allergy." Ironically, in 1988 scrap dental amalgam—fillings that have been removed from the mouth—was declared a hazardous waste product. In an attempt to lead the nation in a more positive direction, California became the first state to ban the use of mercury fillings, effective in 2006.

No Fillings? Not Taking the "Blue Mass"? So What Is Your Exposure?

Mercury is being released into our environment and is polluting our waterways, where the fish we eat are becoming contaminated. The mercury accumulates in the meat of fish and is odorless and invisible.

While some people are certainly more susceptible to mercury poisoning than others, a shocking level of mercury poisoning is being identified right here in America. Once in the human body, mercury acts as a neurotoxin, interfering with the brain and the nervous system, and in young women greatly increases their chance of having babies with birth defects.

Mercury settles into the body by adhering to fat, and because women carry 10 percent more body fat than men, this makes them more prone to mercury poisoning. Children, too, are at great risk because their neurological systems are still developing.

In 2005, the Atlanta-based Centers for Disease Control and Prevention estimated that 1 in 15 U.S. women of reproductive age has a blood mercury level above 5.8 micrograms per liter of blood—a level that could pose a risk to a developing fetus.

A few years ago, a California doctor performed her own study and found that the mercury in her female patients' blood was actually ten times higher than the CDC's average reading; in children the level was sometimes 40 percent higher. Whether this difference reflects the area of the country or there

is some other explanation, the bottom line is that no one is reporting any good news on this subject.

The Harsh Lesson Learned in Japan

If the dangers of mercury were not fully understood early in the twentieth century, they became abundantly clear in Japan during the 1950s. Minamata, located on the western coast of Japan's southernmost island, was downstream of the site of the Chisso Corporation, located in Kumamoto, which by the mid-1930s was manufacturing acetaldehyde, a substance used to produce plastics. From 1932 to 1968, Chisso dumped an estimated twenty-seven tons of mercury compounds into the bay, and it took decades for anyone to realize that the heavy metal had transformed into methylmercury, an organic form that easily enters the food chain.

Given their location by the water, area residents consumed fish regularly, but it was the "dancing cats" in the area that first raised public suspicion. First, the cats began to show signs of erratic behavior. Unfortunately, the symptoms soon began to display themselves in people. Ultimately, over 3,000 residents showed the effects of mercury poisoning, also referred to as Minamata disease, with health abnormalities, dizziness, decreased motor skills, slurred speech, and severe birth defects. Forty-six died.

So What about the Fish?

To date, forty-eight states have issued fishing advisories on mercury. In August 2004, the Environmental Protection Agency (EPA) announced that fish in virtually all U.S. lakes and rivers are contaminated, and in early 2005, a study in the journal *Ecotoxicology* found high levels of mercury in songbirds, salamanders, and other New England wildlife previously thought to be unaffected. This was discouraging news because up until recently it was thought that the mercury poisonings were limited to species that consumed the toxin directly from the water. What they are finding is that when it is airborne—from power plant emissions, for example—the mercury travels

surprisingly far distances and drifts down into leaves and onto plants on land, where it is consumed by sow bugs, centipedes, and other small insects, which then are ingested by birds and other wildlife.

More than thirty years after the alarm was first raised, mercury accumulation in fish remains the chief source of exposure to the toxic metal in the United States. The Food and Drug Administration advises that pregnant women, women who may become pregnant, nursing mothers, and young children avoid shark, swordfish, king mackerel, and tilefish entirely and limit consumption of albacore tuna (canned white tuna and tuna steaks) to 6 ounces (one meal) per week. (These fish are at the top of the food chain and contain higher levels of mercury because they consume smaller fish in polluted streams.) Canned light tuna, shrimp, salmon, pollock, and catfish are said to be tolerable if you eat no more than 12 ounces per week.

While the long-term solution involves reducing the levels of mercury in our environment, there is a short-term solution. If high levels of mercury in the body are detected early, most people can bring their levels down by reducing the amount and the kind of fish they consume. In more severe cases, a chelation process can be helpful in ridding the body of this toxin.

But Where Is It All Coming From?

Coal-fired power plants are the largest single source of mercury emissions in the country, accounting for more than 90,000 pounds of airborne mercury a year—about one-third of the total output. Chlorine plants, which use massive amounts of mercury to extract chlorine from salt, also release a great deal of mercury each year. Facilities that recycle auto scrap are another source of mercury pollution.

Until 2001 factory and power plant emissions were governed by the Clean Air Act, which required plants to have the best available technology in place by 2009. The improvements were projected to lower emissions by 90 percent. The Bush ad-

ministration changed course, removing the power plants from Clean Air Act jurisdiction and proposing the first regulatory effort to cut the emissions with a plan to reduce output by 70 percent within thirteen years. In addition, the EPA also allows the power companies to buy pollution credits instead of reducing emission levels. If polluters are allowed to pay for the right to pollute, then it will worsen emissions in the short term.

And while our first task needs to be to reduce mercury emissions here at home, the issue also needs to be addressed globally as we are learning that to protect our environment we need to take into account the whole world.

So watch your fish consumption, and write to your congresspeople and say we need to reduce our exposure to mercury.

PART FIVE

A Peek at the Future

18

Nanotechnology

A New Frontier

A chapter on nanotechnology seems a fitting way to conclude this book, because this field of science is so new that it's not in regular science books and because nanotechnology offers great promise for the future. From medicine and machinery to energy, the environment, and agriculture, this field could change them all.

So what is nanotechnology? *Nanotechnology* is an umbrella term that describes the science of working in the world of the super tiny—the world of the single atom or the small molecule. To put this in perspective, a nanometer is a billionth of a meter, or a millionth of a millimeter. Scientists predict that within the next fifty years, machines will get smaller and smaller—to the point that thousands of these tiny machines would fit into the period at the end of a sentence.

The science itself involves condensed-matter physics, engineering, molecular biology, and chemistry. Nanoscience has already contributed greatly to many industries—there are products and processes in microprocessor manufacturing, heavy equipment manufacturing, and the aerospace industry. Nanoscience has provided us with the ability to make new catalysts, coatings, paints, and rubber and tire products as well. And you may even be wearing some nanoparticles now. Some forms of sunscreen contain nanoparticles of titanium dioxide, which refracts light. If sunscreen is manufactured with bigger particles, then the sunscreen appears white.

In addition, nanoparticles are being used in specialized items you may have read about. The tennis racquet used by top-ranked player Roger Federer in 2005 was infused with

nanoparticles; they were added to provide a feeling of solidness without adding any weight.

If atoms are nature's building blocks—and are the ingredients of everything from a human to a tree—imagine what might be possible if scientists learn to work at this scale for the good of humanity.

Who Ever Thought of This?

Distinguished physicist Richard Feynman first presented the concept of nanotechnology (but not the term) in a speech he gave to the American Physical Society on December 29, 1959, entitled "There's Plenty of Room at the Bottom." Feynman envisioned the day when we would have the ability to manipulate individual atoms and molecules so that one set of precise tools could be created to make a proportionally smaller set, on down to the needed scale.

The term *nanotechnology* was not used until 1974, when Norio Taniguchi, a professor at Tokyo Science University, introduced it in a paper on the subject. The science got more fully underway when Gerd Binnig and Heinrich Rohrer created the scanning tunneling microscope in 1980, and individual atoms could actually be seen. During the next few years, Dr. Eric Drexler, now regarded as the father of nanotechnology, wrote a book, *Engines of Creation: The Coming Era of Nanotechnology*, that more fully developed the concept.

As we've moved forward from Drexler's original concepts, one thing that has become clear is that nanobots will not be simply scaled-down versions of contemporary robots. The different physics at these scales means that human-made nanodevices will probably bear a much stronger resemblance to nature's nanodevices and will be made from proteins, DNA, and membranes—much like viruses.

Life figured out nanoscience long ago—each one of us has billions of molecular motors crawling around in every cell of our bodies right now. The key is figuring out how to direct human-made ones and keep them going.

It's Working!

One imperative of nanotechnology is that nanodevices must be able to self-assemble. Because these machines will be so small, it will literally take millions of them to do some tasks on a scale that would be helpful to humans.

At Cornell they have created little 4-inch-square blocks that can replicate themselves. The plastic cubes are each 4 inches wide and sliced diagonally into two halves that swivel, and the electromagnets on the cubes turn on and off, allowing them to pick up and release other cubes. A three-cube stack takes a minute to copy itself.

One concern about nanotechnology is that if you build it "well enough," the nanomachine might keep building and building and building unchecked. With this particular device, the Cornell scientists have developed safety features so that the machine cannot go unrestrained. To begin with, the machine relies on power from a base plate, and researchers must also "feed" the robots with new blocks. Another safeguard is that the number of blocks cannot exceed the number of plates.

The Goals of Nanotechnology

Those who foresee a world employing nanotechnology imagine a process that will involve anything from a nanomachine that can create a baseball one atom at a time to another type of nanobot that can speed through one's arteries, cleaning up cholesterol as it tools along. The goal must be to learn how to successfully manipulate material at the molecular and atomic levels, using both chemical and mechanical tools. There have been some successes, and some scientists predict that we could see definite progress within only fifteen to twenty years.

Atoms can now be manipulated, separated, and put back together in different formations. (Atoms and molecules stick together because they have complimentary shapes that lock together or charges that attract.) As millions of these atoms are pieced together by nanomachines, a specific product begins to take shape. To create nanotechnology-produced goods, scientists must learn to further manipulate individual atoms and create "assemblers." Because the scale of this technology is so tiny, trillions of assemblers will be needed to do the work, so

scientists will have to be able to make nanoscopic machines, called *replicators*, that will be programmed to build more assemblers.

The Hope for Nanotechnology

Under President Bill Clinton, the government doubled its investment in research and development of nanotechnology and termed it the new frontier. This work will affect many government agencies: the National Science Foundation, the Department of Defense, the Department of Energy, the National Institutes of Health, the National Aeronautics and Space Administration, and the National Institute of Standards and Technology. Much of the research may take over twenty years, but it could have an amazing effect. This is some of what scientists predict will evolve from this research:

- Nanomaterials are proving to be stronger than steel. By bonding a molecule with a nanoparticle, scientists have created fullerenes, molecules of carbon atoms that when put together form tubular fibers called nanotubes. When those fibers are threaded together and crystallized, they can act like metal but are a hundred times stronger and four times lighter than steel. Large-scale production of such material would change the way not only automobiles are built but airplanes and space shuttles as well.
- Research will lead to stronger fibers. Eventually nanotechnology will be able to copy anything, including diamonds, water, and food. Famine could be eradicated.
- We will see changes in the computer industry. We will eventually be able to store trillions of bytes of information in a structure the size of a sugar cube. And there will be other improvements in computing, sensing, communications, data storage, and display capacities.
- Combat vehicles that are totally nanobot powered could reduce human risk at times of war.

- Research will have a huge impact on medical industry. Consider:
  - Nanobots, controlled by nanocomputers or ultrasound, will be used to manipulate other molecules, destroying cholesterol molecules in arteries, destroying cancer cells, or constructing nerve tissue atom by atom to end paralysis. Patients might ingest the nanobots in fluids, making it a very simple medical process.
  - Nanosurgeons may work at a level a thousand times more precise than the sharpest scalpel, and there should be no scarring.
  - New biomedical solutions for chronic disease will be developed.
  - New drugs and targeted drugs for delivery will be created.
  - Nanoscale medical diagnostic devices will be invented.
  - Though ethics debates will abound, nanotechnology could even change physical appearance by rearranging the atoms of your nose or shifting someone's eye color.

- The research could have a positive impact on the environment.
  - Airborne nanobots could rebuild the thinning ozone layer.
  - Contaminants could be automatically removed from water sources, and oil spills could be cleaned up instantly. Nanotubes with fingers 50,000 times thinner than a human hair would manipulate the atoms in an oil spill to render it harmless.
  - Nanotechnology will pollute less, and we will reduce our dependence on nonrenewable resources.

- Many resources could be constructed by nanomachines. In the energy field, atoms bonded together could create a machine that converts water to hydrogen using sunlight to create a limitless energy source.

- People in hazardous environments will have greater protection. Clothing will constantly monitor physiological vital

signs, warn about exposure to chemicals, adjust for environment stresses, provide camouflage that matches changing background and lighting conditions, and even provide first-aid casualty response.

- Space exploration will expand. From unmanned explorations into deep space to a big improvement in materials for spacecraft and reentry vehicles, nanotechnology will be a huge benefit here.

Potential Risks

In the wish-I-hadn't-said-that category, Dr. Eric Drexler, the father of nanotechnology, almost certainly regrets his early speculation that one of the slight risks of nanotechnology is that the self-replicating feature could lead to a situation in which nanobots run amok and take over the world—what is now termed global ecophagy (literally, "eating the environment"). A variation of this fear is the "green goo" theory, in which nanobiotechnology creates a self-replicating nanomachine that consumes all organic particles, creating a slime-like nonliving organic mass.

The Risk of Gray Goo

In a worst-case scenario, all of the matter in the universe could be turned into goo, killing the universe's residents. (Goo means a large mass of replicating nanomachines lacking large-scale structure; gray goo refers to runaway nanobot self-replicators; green goo refers to an organic replicator as described above.) In Michael Crichton's recent novel *Prey*, a company in Nevada accidentally/purposely releases self-assembling nanobots into the desert; they quickly replicate and evolve and threaten all the human protagonists.

According to Chris Phoenix, director of research of the Center for Responsible Technology, runaway replicators would be only the product of a deliberate and difficult engineering process, and not an accident. Drexler notes that the scaremongering detracts from the promise of nanotechnology. But with

the risks in mind, scientists are programming nanomachines to rely on an energy source or to stop reproducing after a certain number of generations. That way a nanomachine could not go unchecked.

Bacteria have never taken over the world, so nanoparticles probably won't either.

Poison and Toxicity

Some of the most promising nanoscale materials have been found to be harmful in certain situations, so scientists will want to proceed cautiously here. A high concentration of carbon nanotubes can fatally clog the lungs of rats, and fullerenes (spherical carbon cages just a few nanometers across) tend to accumulate in fish brains.

Nanoparticles in drinking water could be dangerous to humans or animals, and this new class of nanosubstances will need to undergo additional safety testing.

Weapons

Weaponization would obviously be a very negative use of nanotechnology. While advanced nanomaterials have applications for improving existing weapons and military hardware, currently nanotechnology could not be weaponized in any way that is more dangerous than what is already available to us via genetic engineering.

While one might envision a nanomachine that could eat rubber and disable vehicles by destroying the tires, at this point, there are many other ways to accomplish greater destruction.

Ethical Issues and Privacy

Nanotechnology could be used to create highly undetectable surveillance devices—molecular-sized microphones, cameras, and homing beacons are all possible.

It is already possible to use a rice grain–sized radio frequency identification for calling up medical records, but this brings up the issue of privacy.

So What's a Nanotube Good For?

In January of 2006, fourteen years after the discovery of the pencil-shaped molecules called carbon nanotubes, scientists are finding a variety of applications for them. Nanotubes are 9 times stronger than steel and can transmit 1,000 times more electrical current than copper, but they are difficult to manipulate because each tube is a tiny fraction the width of the period at the end of this sentence.

A group at the University of Texas, under Ray Baughman, has learned to weave nanotubes into a useful material. Using Australian wool spinners, researchers have developed a method for twisting the tubes into long fibers, and they have created sheets of nanotubes so thin that an acre of the material weighs just a quarter of a pound.

Two teams are working on medical applications of nanotubes. Because the human body can absorb carbon, scientists at Stanford University have created cancer-killing nanotubes designed to invade tumor cells. At the University of California, Riverside, they are experimenting with ways to use nanotubes to help heal broken bones.

Index

Note: Page numbers followed by *f* indicate an illustration.